AUSTRALIAN ENVIRONMENTS

The Great Barrier Reef Marine Park

ISBN: 978-1-7641194-9-8 (Paperback)

First edition: May 2026
A CONSERVATION CLUB PUBLICATION
www.conservationclub.com.au

NATIONAL LIBRARY OF AUSTRALIA

A catalogue record for this book is available from the National Library of Australia

In the spirit of reconciliation, Conservation Club acknowledges the Traditional Custodians of Country throughout Australia and their connection to land, sea, and community. We pay our respects to their Elders, past and present, and extend that respect to all Aboriginal and Torres Strait Islander peoples.

To my grandchildren: Samuel, Jameson, Harry, and Audrey

..in the hope that they will enjoy sharing this journey of discovery ..

"And, above all, watch with glittering eyes the whole world around you because the greatest secrets are always hidden in the most unlikely places. Those who don't believe in magic will never find it."

Roald Dahl

CONTENTS

REWARDS:

Conservation Club offers rewards (such as free downloads of colouring sheets and button badges) that can be earned by answering the 20 Question Quiz on page 147 and submitting one coloured-in image with the colours being truly representative of the species concerned. We encourage keen eyes, writing, recording, and reporting species sightings.

- **Buy one of our pocket notebooks! Designed to fit in a child's pocket, this blank notebook is approximately A7 size and an invaluable tool for recording your observations of the environment you are exploring. It is really important to note the date, time of day, weather conditions and details of any sightings you may be lucky enough to experience! If the species is listed on the NCA, EPBC, and/or IUCN Red List, please consider reporting your observations to an entity such as iNaturalist or the Atlas of Living Australia! Citizen scientists are making great contributions to our understanding of these environments – you could be one of them!**
- **It is totally okay for you to guide your youngsters and help them with this task (we suggest using the form at the back of the book as a guide before downloading the questions)!**

Email: conservationclubau@gmail.com for more information about how to apply for these rewards. The quiz questions are on the last page of this book. You can also join our mailing list to keep up to date with new releases.

Author's notes:

- Words and phrases shown within the book text in **BOLD** appear in the extensive glossary/word lists at the back of this book.
- The whole point of the colouring pages is to encourage your young **conservationists** to engage with the book, tweak their observational skills, do a little bit of research, with the help of an adult if needed, and find out what colours are appropriate for the **species** concerned!
- Some of the words and language used may well be beyond what your young **conservationist** can understand, but it is suggested that you use the glossary to explain those words/phrases in simpler terms and at a level appropriate for your young wildlife warrior. They will ask for more information when they are ready!
- You can also choose to share the facts that you feel your young wildlife warriors are able to understand—you do not have to use them all! Just pick a few based on your child's age and comprehension skills… You can always come back to the text later to expand their knowledge and vocabulary as they learn.

COLOURING RECOMMENDATIONS:

We recommend using colour pencils rather than texters or felt pens to ensure the information pages remain readable. Texters can bleed through the colouring side, affecting the readability of the fun facts!

We encourage adults and youngsters to find coloured images or photos online to guide the colouring. Familiarising both you and them will help you recognise the **species** if you sight them!

INTRODUCTION

The Great Barrier Reef Marine Park

In these colouring pages, we focus on the environment of **The Great Barrier Reef Marine Park** and explore some of the animals and plants that call it home. Some of the featured **species** are **critically endangered, endangered,** or **vulnerable;** the information for each species also includes this classification, where available. **The Great Barrier Reef** is located in the **Coral Sea**, off the coast of Queensland**,** Australia**,** separated from the coast by a channel 160km wide in places and over 61m deep. The Great Barrier Reef can be seen from outer space and is the world's biggest single structure made by living **organisms**. This reef structure is composed of and continues to be built by billions of tiny **organisms**, known as **coral polyps**. The hard, stony corals, or **scleractinians**, are the foundation of coral reefs, building themselves a hard skeleton of **calcium carbonate**.

The Great Barrier Reef is the world's largest coral reef **ecosystem** and one of just a few World Heritage Areas that include **marine** and **coastal** values that meet all four natural World Heritage criteria. The Great Barrier Reef Marine Park (GBRMP) is also the world's largest **marine** protected area (344,400km^2; more than 2300km in length and several hundred kilometres wide in some parts). The GBRMP encompasses a large number of **habitats** (e.g., more than 2900 reefs and about 900 islands and **coral cays**, comprising 30 reef and 40 non-reef **bioregions**) due to its size. This size underlies the vast **diversity** of **marine** life in the region and its **proximity** to the **global** centre of **marine biodiversity**. About 99.3% of the World Heritage Area is within the GBRMP, the remaining 0.7% being Queensland waters and islands, most of which are contained within a complementary system of **marine** parks managed by the Queensland Government.

The Great Barrier Reef Marine Park Act also established the Great Barrier Reef Marine Park Authority ('the Authority'). The Authority's goal is: *'To provide for the protection, wise use, understanding and enjoyment of the Great Barrier Reef in perpetuity through the care and development of the Great Barrier Reef Marine Park'*. The stated aims of the Authority include protecting the natural qualities of the Great Barrier Reef while providing for reasonable use of the Reef Region, and minimising regulation of, and interference in, human activities, consistent with meeting the goal and other aims of the Authority. Consistent with these obligations, the Authority is responsible for conserving **species** of high **conservation** value (e.g. rare, threatened) in the GBRMP. This is achieved by managing human activities and their associated impacts in the GBRMP, including current and predicted future activities. To the extent that it is consistent with protecting the natural values of the Great Barrier Reef, the Authority provides for ecologically sustainable use of the Marine Park.

The way in which the Great Barrier Reef Marine Park Authority (GBRMPA) manages for the conservation of species in the GBRMP is determined mainly by legislative instruments and policy decisions, which are supported by education and enforcement. Given the **migratory** nature of many **species**, the GBRMPA works closely with other Australian and Queensland Government agencies to ensure complementary management approaches, to the extent possible.

Environmental pressures on the reef and its **ecosystem** include **runoff** of human-made **pollutants**, climate change accompanied by mass coral bleaching, dumping of dredging sludge and cyclic population outbreaks of the crown-of-thorns starfish. According to a study published in October 2012 by the Proceedings of the National Academy of Sciences, the reef has lost more than half its coral cover since 1985, a finding reaffirmed by a 2020 study, which found over half of the reef's coral cover to have been lost between 1995 and 2017, with the effects of a widespread 2020 bleaching event not yet quantified.

The Australian Institute of Marine Science conducts annual surveys of the Great Barrier Reef's status, and the 2022 report showed the greatest recovery in 36 years. It is mainly due to the regrowth of two-thirds of the reef by the fast-growing *Acropora* coral, which is the **dominant** coral.

The Great Barrier Reef has long been known to and used by the Aboriginal Australian and Torres Strait Islander peoples, and is an important part of local groups' cultures and spirituality.

The first European to sight the Great Barrier Reef was James Cook in 1770, who sailed and mapped the east coast of Australia. On 11 June 1770, Cook's ship, HMS Endeavour, ran aground on a shoal south of the present-day location of Cooktown, requiring seven weeks to repair. It was Matthew Flinders who named the Great Barrier Reef after his more detailed mapping of it in 1802. Flinders used various terms to describe the reefs comprising what we now call the Great Barrier Reef including "great reef", for one such reef, "barrier reef", for any reef preventing a sailing vessel in, or waves from, the open sea, from reaching the coast, and "barrier reefs", for the collection of such reefs.

The Great Barrier Reef's development history is **complex**; after Queensland drifted into **tropical** waters, it was largely shaped by reef growth and decline as sea levels rose and fell.

Reefs can increase in diameter by 1 to 3cm per year, and grow vertically anywhere from 1 to 25cm per year; however, they grow only above a depth of 150m due to their need for sunlight, and cannot grow above sea level.

When Queensland edged into **tropical** waters 24 million years ago, some coral grew, but a **sedimentation** regimen quickly developed with erosion of the Great Dividing Range; creating **river deltas**, **oozes** and **turbidities**, unsuitable conditions for coral growth. Then, 10 million years ago, the sea level dropped significantly, further enabling **sedimentation**.

From 20,000 to 6,000 years ago, sea levels rose steadily worldwide. As they rose, the corals could then grow higher on the newly submerged maritime margins of the hills of the coastal plain. By around 13,000 years ago, the sea level was only 60m lower than the present day, and corals began to surround the hills of the coastal plain, which were, by then, **continental islands**. As sea levels rose further, most of the **continental islands** were **submerged**. The corals could then overgrow the **submerged** hills to form the present **cays** and reefs. Sea level here has not risen significantly in the last 6,000 years.

WILDLIFE IN THE GREAT BARRIER REEF MARINE PARK

The Great Barrier Reef supports an extraordinary **diversity** of life, including many **vulnerable** or **endangered species**, some of which may be **endemic** to the reef system. Here are just some of the statistics:

- 30 species of **cetaceans**
- 1500+ fish species
- 17 species of sea snake
- 6 species of sea turtles
- 15 species of seagrasses
- 125 species of shark, stingray, skates or chimaera
- 5000 species of mollusc
- 49 species of pipefish, 9 species of seahorse
- 7 species of frogs (islands)
- 215 species of birds (including 22 seabirds and 32 shorebirds)
- 2195 species of plants (islands)
- 330+ species of **ascidians**
- 300-500 species of **bryozoans**
- 400 coral species (hard and soft)
- 500 species of **marine algae**

For more information, you can search Great Barrier Reef and find multiple sources of information, some much more detailed than provided here. Wikipedia is a great source of information!

<u>**It should be noted that the Great Barrier Reef Marine Park boundaries include some continental islands and coastal areas extending inland of the Great Barrier Reef itself.</u>

DID YOU KNOW?

Australia is the sixth largest country on Earth and the only island/country that covers an entire **continent**! Because of its **isolation** from other **continents**, many of Australia's animal and plant **species** are found nowhere else. Australia also has a wide variety of **environments**, many of which are threatened for various reasons, including climate change, **extreme weather events**, and human impacts.

Around 80% of Australia's **native species** are **endemic,** and many are likely to become **extinct** within our lifetimes. Conservation Club AU focuses on raising awareness of **species** and **environments** in our 'backyards,' which we may or may not be familiar with. Based on **fossil** records, the natural **extinction** rate is approximately one **species** per one million **species** yearly. Scientists estimate that around 8.7 million **species** of plants and animals currently live on Earth. They are racing against time to catalogue Earth's **biodiversity** as **extinctions** continue. The Australian **continent** has seen more **mammal extinctions** than any other, along with **significant** losses of frogs, lizards, fish, insects, and plants. Many of these **extinctions** are recent. In Australia, there have been 21 likely **extinctions** of animals and four of plants, averaging about four per **decade**. Out of the 23 **extinctions** for which causes have been identified, 17 (73%) were primarily due to **invasive species**. Focusing just on animals, 17 out of 19 (89%) **extinctions** were mainly caused by **invasive species**. The four plant **extinctions** are primarily attributed to **habitat destruction** or **degradation**. The **primary culprits** in recent **extinctions** include the ***chytrid*** **fungus** (affecting **amphibians**)**, which has wiped out six **species**, and the wolf snake, which is responsible for the loss of four **species**. Other **significant** impacts include cats, foxes, black rats, brown trout, and a ***trypanosome*** blood **parasite**. Although **habitat** loss is often **cited** as the **primary** cause of **extinctions**, this does not hold true for recent **extinctions** in Australia. Earlier conclusions linked many plant **extinctions** to **habitat** loss. Still, increased efforts in searching have led to the rediscovery of many of these plants. The current data indicate that **invasive species** have been the leading cause of most **extinctions** in Australia since **European settlement.**

It has been extremely difficult to decide on which species to include and which species to exclude for this book! I have tried to include the species that are of particular concern, as well as those I think are most likely to make interesting colouring-in pages. As previously stated, the Great Barrier Reef Marine Park is a vast geographical area, and many of the species, whilst they may appear on the IUCN Red List, are extremely difficult to assess realistically!

It should also be noted that numerous corals are on the IUCN Red List; however, correctly identifying coral species is a relatively exacting task, with many very similar to each other and some species with multiple different growth forms. Therefore, despite there being many different corals in the illustrations, only two have been specifically illustrated for the purposes of this book.

LET'S EXPLORE!

Kids in the 8 to 14-year-old age group are generally interested in spending time with their families exploring the great outdoors on weekend walks, camping getaways, etc. What a great way to spend time with your kids! It is so much better than spending hours on electronic media and becoming **sedentary**, both of which can lead to behavioural problems and sleep issues!

Kids also love to be part of a club and start collecting things they might be interested in, such as pebbles and shells. In my opinion, our kids are generally far more aware of **conservation** and **environmental** issues than previous generations were. This, therefore, presents an opportunity to educate, to get them interested, active, **motivated,** and involved in the **environment** and the **conservation** of **species** so that they will, hopefully, be able to work to preserve these animals so that they can show or tell their own children about them in years to come.

PLEASE.... GET OUT THERE and encourage these young, curious minds to engage with their local environments and the critters they can find there!

NATIONAL & MARINE PARKS

Numerous Queensland National Parks are located within the Great Barrier Reef Marine Park (GBRMP), including the islands and coastal areas of the Wet Tropics World Heritage Area and the islands in the Whitsundays. Examples include Hinchinbrook Island National Park, Magnetic Island National Park, Paluma Range National Park, and Whitsunday Islands National Park. Most of the islands are within the World Heritage Area and are managed jointly by the GBRMPA and Queensland Parks and Wildlife Service. North West, Lady Musgrave and Mast Head islands, all coral cays in the Capricorn Bunker Group, have campgrounds (no water, composting toilets) that you can visit after obtaining permits from the National Parks and Wildlife Service. Although this requires significant planning, it is well worth the effort. Boat transfers are available from various locations depending on which island you are visiting, for example, from Yeppoon (Rosslyn Bay), Gladstone, Bundaberg and the Town of 1770 or for groups via private charter. Heron, Lady Elliot, and Wilson islands have some facilities; however, Wilson Island is adults-only.

Marine parks

- Great Barrier Reef Marine Park (GBRMP)
  This is a Commonwealth marine park that covers the vast majority of the reef **ecosystem** (approx. 344,400 km^2) and is managed by the Great Barrier Reef Marine Park Authority (GBRMPA, also known as the "Reef Authority").
- Great Barrier Reef Coast Marine Park
  This is a Queensland State marine park that adjoins the GBRMP in **coastal tidal** lands and waters, including most river and **creek** mouths. It uses zoning that is generally complementary to the GBRMP.

Coastal and mainland parks

- Annan River (Yuku Baja-Muliku) National Park
- Apudthama National Park (Cape York Peninsula Aboriginal Land)
- Bowling Green Bay National Park
- **Conway National Park**
- Daintree National Park (Cape York Peninsula Aboriginal Land)
- Girramay National Park
- Paluma Range National Park
- Girringun National Park

Island and cay national parks

- Barnard Island Group National Park
- Brampton Islands National Park
- Broad Sound Islands National Park
- Brook Islands National Park
- Capricornia Cays National Park
- Family Islands National Park
- Fitzroy Island National Park
- Frankland Group National Park
- Gloucester Islands National Park
- Green Island National Park
- Hinchinbrook Island National Park
- Holbourne Island National Park
- Hope Islands National Park (Cape York Peninsula Aboriginal Land)
- Keppel Bay Islands National Park
- Lindeman Islands National Park
- Lizard Island National Park
- Magnetic Island National Park
- Michaelmas and Upolu Cays National Park
- Molle Islands National Park
- Orpheus Island National Park
- Repulse Islands National Park
- Swain Reefs National Park
- Whitsunday Islands National Park

Some background information about this book:

*The **species** in this book occur in the Great Barrier Reef Marine Park. Some of the **species** illustrated are **Critically Endangered, Endangered, Vulnerable, Near Threatened or of Least Concern** on the NCA and EPBC (SPRAT) lists.*

*On the IUCN Red List.[1], despite this being extensive, more than 172,620 **species** have been assessed at the time of writing this book. However, of the goal of 260,000 **species**, there are still 87,380 **species** remaining to be evaluated, an incredible achievement, but, for some **species**, due to the lack of up-to-date scientific data, this list is, at this point in time, not complete. The list is constantly updated, with the goal of reassessing 142,000 **species**. More information can be found at https://www.iucnredlist.org/*

The following tables provide information on how the species are listed in the NCA, EPBC (within Australia), and IUCN lists, the categories they are listed under, and a breakdown of the categories currently available for assessment. The EPBC and NCA listings are not always available for all species.

Younger readers may be unable to navigate this information on their own, so it is provided here to enable parents or caregivers to relay it to those young, curious minds!

I firmly believe that the age groups this book is aimed at (8 to 16 years) are curious and generally more informed than previous generations. They are aware that even one single voice can make a difference. This is evidenced by the number of young people taking up a cause. They recognise that, even one voice, one person, can make a difference, as evidenced by the number of young people taking up a cause.

Now, with both individuals and the 'global hive mind' enabled by improved communication systems, the internet, data sharing, and, more recently, AI, I believe this is bringing these individuals and their causes into clear focus, and they demand to be heard regardless of age!

The book is likely to appeal to a much broader audience than just the age group mentioned, as the illustrations and information are also appropriate for adults. Most importantly, we all need to be much more aware of our surroundings and report our observations to an entity such as SPRAT.

I sincerely hope that you, the Wildlife Warriors for the Great Barrier Reef Marine Park environment, will enjoy this book as much as I have enjoyed writing and illustrating it!

Elysia Rainwood

NCA – Status under the Nature Conservation Act 1992 (QLD legislation) [21]	
EX	Extinct
PE	Extinct in the wild
CR	Critically endangered
E	Endangered
V	Vulnerable
NT	Near threatened
C	Least concern
SL	Special least concern
I	International
P	Prohibited

EPBC – Status under the Environment Protection and Biodiversity Conservation Act 1999 [2]	
EX	Extinct
XW	Extinct in the wild
CE	Critically endangered
E	Endangered
V	Vulnerable
CD	Conservation dependent

IUCN – Status on the IUCN Red List [1]	
EX	Extinct
EW	Extinct in the wild
CR	Critically endangered
EN	Endangered
VU	Vulnerable
NT	Near threatened
LC	Least concern
DD	Data deficient
NE	Not evaluated

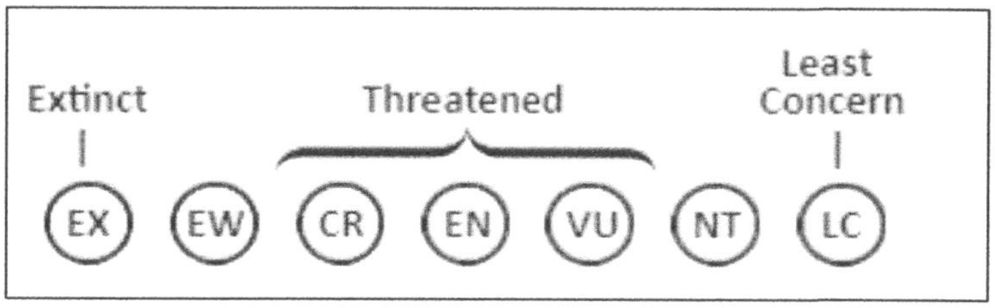

Where the species are on the **IUCN Red List,** the category is outlined in bold and grey in colour!

IUCN categories explained: [1]

CE = Critically Endangered:....... Facing an extremely high risk of extinction in the wild.

EN = Endangered:...................... There is a very high risk of extinction due to rapid population declines of 50 to more than 70% over the previous 10 years, a current population size of fewer than 250 individuals.

V = Vulnerable:.......................... Being threatened with extinction unless the circumstances that are threatening its survival and reproduction improve.

- Sometimes the species on the IUCN Red List have not been listed simply because there has not been sufficient research to assess.
- In these cases, they are regarded as DD = Data Deficient **OR** NE = Not Evaluated.
- That does not mean that they are not critically endangered, endangered, vulnerable, threatened or any other classification... It just means that they have not been assessed!

Where on earth is the Great Barrier Reef Marine Park found?

Here is a map that helps to delineate the area of the Great Barrier Reef Marine Park:

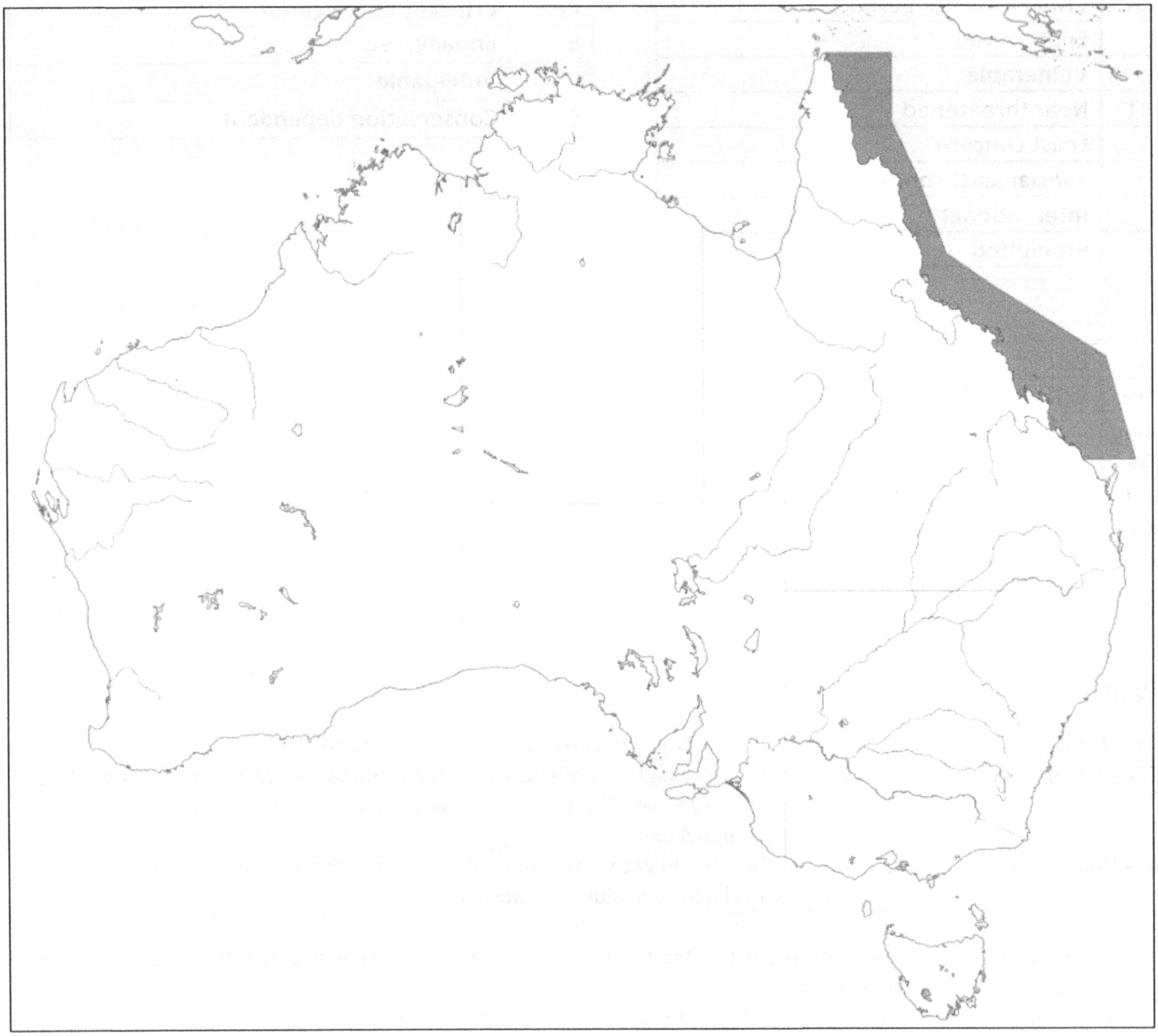

BONY FISH EXTERNAL ANATOMY

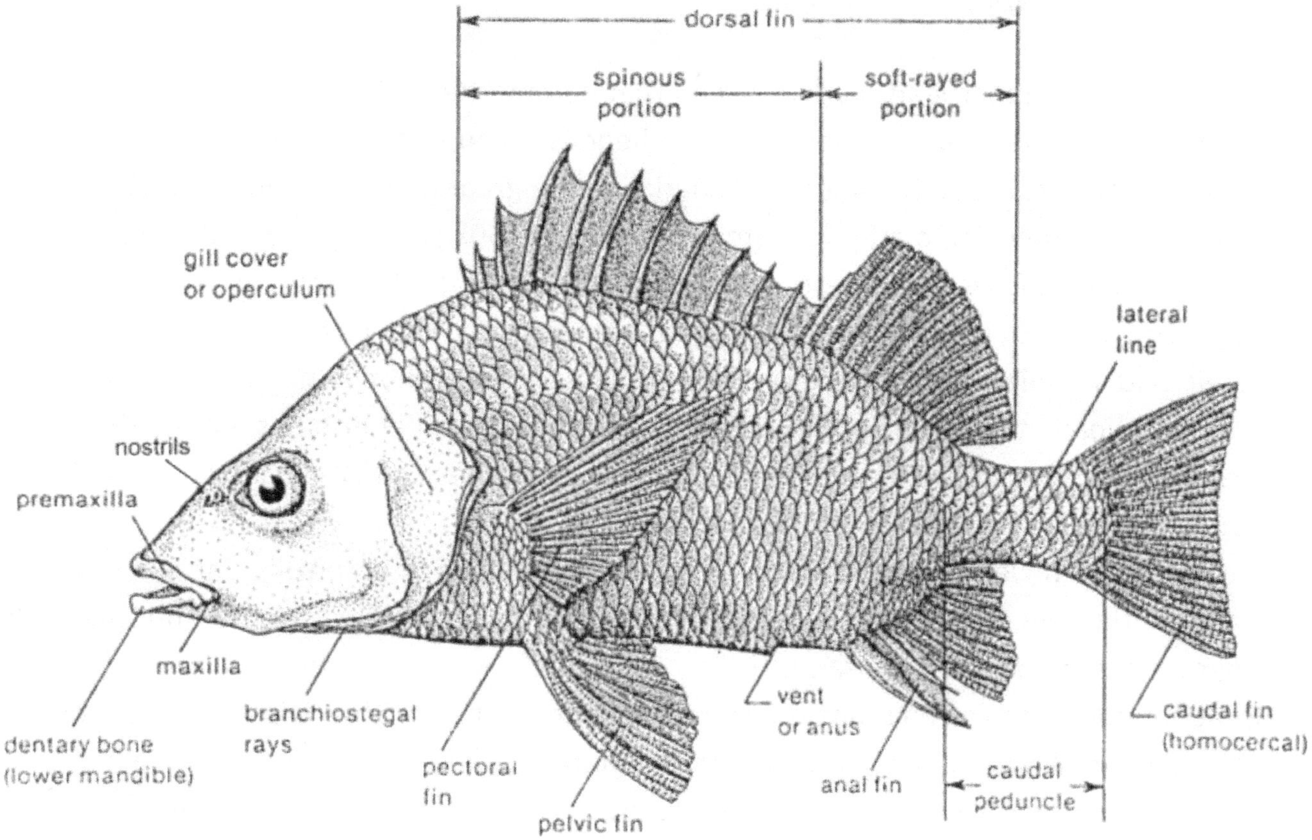

BASIC BIRD EXTERNAL ANATOMY

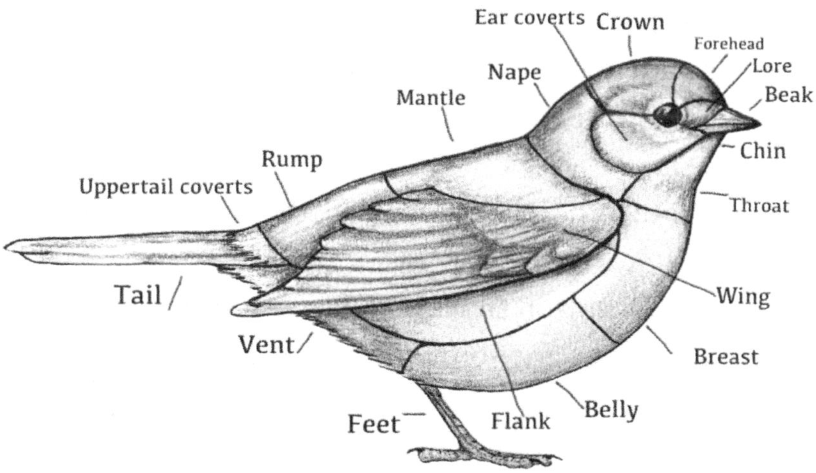

Great hammerhead and scalloped hammerhead sharks

Sphyrna lewini and *Sphyrna mokarran*

- The scalloped hammerhead shark and the great hammerhead shark are two of nine hammerhead shark **species.** Their T-shaped head is called a **cephalofoil**.
- Comparing the scalloped hammerhead shark and the great hammerhead shark, the scalloped hammerhead grows up to around 4.3m in length and 150kg in weight, and the great hammerhead, which is the largest of all the hammerhead sharks, grows up to around 6m in length and 580kg in weight. Whilst the great hammerhead shark's head has a straight front margin and a shallow notch in the centre of the head, the scalloped hammerhead shark has an indentation in the middle of the expanded head, and smaller notches around the head.
- Because their eyes are on the very ends of their heads, they can see in a full circle (360°) around them—above, below, and to the sides—all at once!
- Hammerheads have special sensory organs called **ampullae of Lorenzini**. These let them sense the tiny electrical signals given off by other animals, allowing them to find snacks like stingrays hidden beneath the sand.
- Great hammerheads often swim at a slight angle on their side. This helps them use their extra-tall **dorsal** fin for lift, which saves them energy.
- Unlike many other fish, hammerhead sharks give birth to live babies called pups.
- These two hammerhead species generally have between 12 and 40 pups at a single time.
- From the moment they are born, pups are on their own and have to learn to hunt and swim immediately.
- Their favourite food is stingrays. They actually use their wide heads like tools to pin down stingrays on the seafloor so they can't swim away.
- They are mostly **immune to** the **venomous** stings of rays. Scientists often find stingray barbs stuck in a hammerhead's mouth and head that don't seem to bother them at all.
- While they love rays, they also eat fish, squid, crabs, and even other smaller sharks.
- While many sharks prefer to be alone, hammerheads often hang out in big groups called **schools** during the day.
- Even though they hang out together during the day, they usually split up at night to hunt for food by themselves.
- Hammerheads spend so much time in shallow, sunny water that they can actually get a tan, turning from light grey to a darker grey-brown.

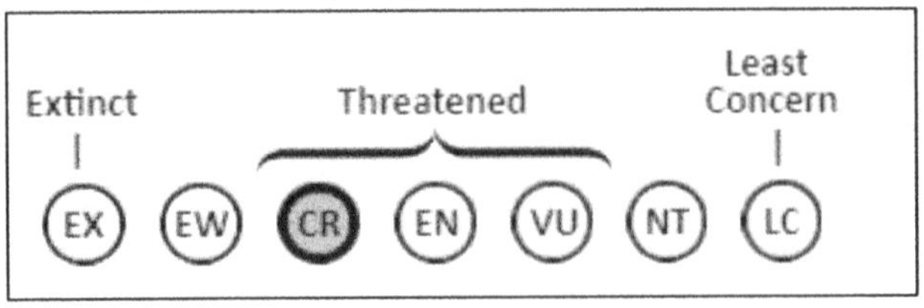

[1]

NCA List = Endangered
EPBC List = E = Endangered

Orange-spotted filefish, harlequin filefish

Oxymonacanthus longirostris

- These fish absorb the scent of the *Acropora* coral they eat, allowing them to hide from **predators** at night by smelling exactly like the coral they are nestled in.
- At night, they anchor themselves into coral branches in a head-down position to stay safe and **camouflaged** while they sleep.
- Their blue-green skin with orange-yellow spots perfectly matches the appearance of the coral branches and **polyps** they inhabit, making them nearly invisible to the naked eye at a distance.
- Like other filefish, they have rough, non-overlapping scales with small spikes, giving their skin a texture similar to sandpaper or a file—which is how they got their name.
- In the wild, they exclusively feed on the **polyps** of *Acropora* corals. Due to this **specialisation**, they often starve if their coral habitat dies, refusing to switch diets.
- They have a black spot on their tail, but it disappears when they flare their **caudal fin**.
- Instead of having traditional **pelvic fins**, they have a specialised "pelvic flap".
- Males have a bright orange spot on their **pelvic** flap. During **courtship**, they can rapidly move this flap, creating a flickering effect that looks like a flashing orange light to attract a female.
- These fish are typically found in **heterosexual** pairs that swim together in **synchronised** motion, often staying together for life
- While they are a **coveted**, beautiful **aquarium** fish, they are notoriously difficult to keep. While many die in captivity, some enthusiasts have managed to train them to eat frozen foods such as brine shrimp.
- Found in the Indo-Pacific, often in pairs, living on coral reefs at depths of 1 to 30 metres.
- Relying on their **camouflaged** appearance and coral **habitat**, they are not fast swimmers and use their **pectoral** and **dorsal fins** for slow, coordinated movement.
- They are **diurnal**, spending the day feeding and hiding in coral, making them a very shy species in new environments.

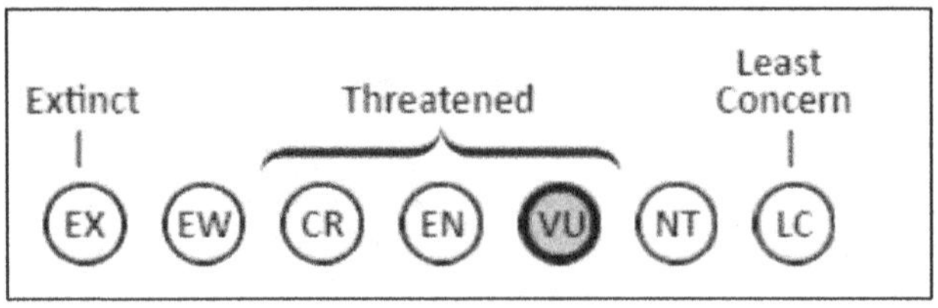

[1]

EPBC List = Not listed
NCA = Not listed

Whitespotted eagle ray

Aetobatus ocellatus

- Unlike bottom-dwelling rays, these eagle rays swim in the open ocean or near reefs, moving by flapping their powerful **pectoral fins**, **mimicking** the flight of a bird.
- They are known to leap entirely out of the water to escape **predators** like sharks, sometimes reaching heights of up to 2m. Occasionally, their leaping behaviour leads them to land on boats, and in some instances, this has caused injury to passengers.
- They have a **distinctive**, flat, duck-like snout, which they use to dig through sandy seabeds to find buried **molluscs** and **crustaceans**. They do not have sharp teeth. Instead, they use a single row of **chevron**-shaped teeth to crush through hard-shelled prey like oysters, clams, and sea urchins. When feeding in the sand, they often disappear in a cloud of dust that they blow out of their **gills**.
- The **dorsal** side is dark grey to blue, covered in many small white spots or rings (**ocelli**), and has a pale underside for countershading **camouflage**.
- While their disc (**wingspan**) is typically around 1.5–3 meters, they can have very long, whip-like tails, allowing them to reach a total length of over 8 meters; adults may weigh over 200kg! Females are larger than males; males have **claspers** (modified pelvic fins) used during mating.
- They have 2 to 6 **venomous barbed** spines at the base of their tail, used for **defence** against sharks.
- They are **ovoviviparous**, meaning the eggs develop and hatch inside the mother. The pups are born as miniature versions of the adults after a 12-month **gestation**.
- Females typically give birth to only 1–4 pups at a time. Researchers have found evidence that, in rare instances, these rays can undergo **parthenogenesis** (a "virgin birth"), in which the egg develops without **fertilisation**.
- They have been observed visiting cleaning stations where wrasses and other fish pick **parasites** off them.
- While sometimes solitary, they often form "schools" or "fevers" to swim together in large numbers.
- They are sometimes observed to show **curiosity** toward divers or snorkelers, circling back to observe human visitors.
- Due to their slow **reproductive** rate and accidental capture in fishing gear (**bycatch**), they have been reclassified, with some populations considered **endangered** and having declined by 50–79% in some areas over the last few decades.

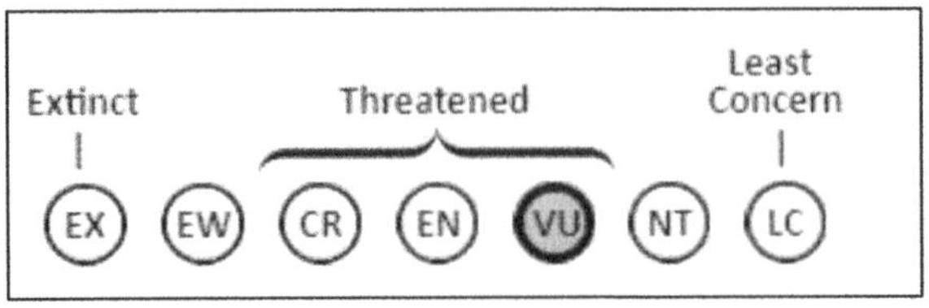

[1]

EPBC List = Not listed
NCA = Not listed

Indo-Pacific sailfish

Istiophorus platypterus

- The Indo-Pacific sailfish is arguably the fastest fish in the ocean, capable of high-speed hunting using its spear-like bill. They are instantly recognisable by their massive **dorsal** fin, which acts like a sailboat's sail. These fish, which can grow over 3m long, hunt in groups and can change colour when excited.
- They can reach speeds up to 70 mph (110 kph), making them one of the fastest animals in the ocean. Unlike many fish, they have scales **embedded** in their skin, helping them remain incredibly smooth and **hydrodynamic**.
- The first dorsal fin is taller than its body and runs nearly its entire length. It remains folded down during high-speed swimming to reduce drag, but is raised for stability to **herd** prey, or during "sail-raising" behaviour to possibly regulate body temperature.
- Sailfish are among the fastest-growing fish; they can reach lengths of 1.2m to 1.5m in just one year.
- Their bodies can change colour based on their level of excitement or **fatigue**, which is controlled by their nervous system. They often "light up" with bright neon-blue spots or **vertical** bars when hunting to confuse **prey**.
- They have a **specialised** ability to heat their brain and eyes through a network of blood vessels, which helps them see clearly and maintain sharp reaction times while hunting in colder, deeper waters.
- They use their long, round **bill** to slash through schools of fish like sardines or mackerel, stunning or injuring them before circling back to eat them.
- Sailfish are known for "**herding**" **baitfish** into tight balls, using their large sails to create a "wall" that prevents prey from escaping.
- Beyond schooling fish, they are **opportunistic** hunters that also **consume** squid, octopus, and **crustaceans**.
- In the wild, they typically live about 13 to 15 years, though many caught and released by **anglers** have a shorter average **lifespan** of 4 to 7 years.
- A single large female can release up to 4.5 million eggs, which float near the surface during a **spawning** session to ensure survival in the open ocean.
- When they first **hatch**, they are tiny (about 0.3cm) and lack the **characteristic** long **bill**. The **bill** begins to elongate when they reach about 0.6cm.

[1]

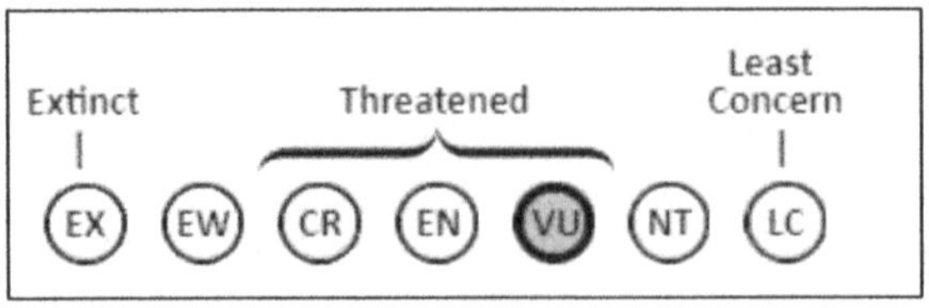

EPBC List = Not listed
NCA = Not listed

Butterflyfish

Family = Chaetodontidae. Genus = Chaetodon

Illustrated here are six species, all of which have been assessed on the IUCN Redlist:

- **Chevron butterflyfish** *(Chaetodon trifascialis)* – IUCN Redlist = NT
- **Rainford's or Gold-barred butterflyfish** *(Chaetodon rainfordi)* – IUCN Redlist = NT
- **Merten's butterflyfish** *(Chaetodon mertensii)* – IUCN Redlist = LC
- **Ornate Butterflyfish** *(Chaetodon ornatissimus)* – IUCN Redlist = LC
- **Saddle butterflyfish** *(Chaetodon ephippium)* – IUCN Redlist = LC
- **Pinstripe butterflyfish** *(Chaetodon lunulatus)*– IUCN Redlist = LC

Can you tell which are which from their size, shape of body, shape of fins, markings, etcetera?

- Butterflyfish are brightly coloured, thin-bodied reef fish, known for forming lifelong **monogamous** pairs and acting as specialised **corallivores** that feed on coral **polyps** during the day. While most are 12 to 22cm long, some species, like the lined and saddle butterflyfish, can grow to 30cm. They generally live for 5–10 years in the wild, with some even longer.
- Their family name, *Chaetodontidae*, means "bristle-tooth" in Greek, which describes their **specialised**, tiny, comb-like teeth. Most butterflyfish are **specialised** eaters, feeding **specifically** on coral **polyps** and sea anemones. They use their small, **specialised** mouths and often long, pointed snouts to carefully pick small **invertebrates**, **zooplankton**, and coral polyps from within small crevices. Some are "obligate corallivores," meaning they eat only live coral **polyps**, making them vital indicators of reef health for scientists.
- Many species of butterflyfish live in male/female pairs that stay together for several years, sometimes for life.
- These **monogamous**, **diurnal** fish primarily **inhabit** shallow, **tropical** waters and are renowned for their **precision** swimming within coral crevices.
- Many butterflyfish have a dark, **vertical** stripe that hides their actual eye, along with a **prominent**, dark eyespot on their rear body, which tricks **predators** into attacking the tail end instead of the head, giving the fish a chance to escape in the opposite direction.
- They are also known for having a **unique** armour-plated **larval stage**. When young, butterflyfish are in a "**tholichthys**" stage, where their bodies are covered in large, bony plates that extend from the head. They lose these plates as they **mature**.

.... Continued on page 22.......

Butterflyfish (continued)

Chaetodon sp.

- Though mostly **diurnal** (active by day), their appearance changes dramatically at night. Their bright yellow and white colours often fade or develop dark patterns, helping them blend into reef shadows while they sleep.
- They are highly **territorial**, often staking out specific coral heads.
- Some juvenile butterflyfish act as "cleaner fish," removing **parasites** from other fish.
- When **spawning**, the pair often rises toward the water's surface, with the male nudging the female's belly before she releases eggs.
- Butterflyfish can make noises to communicate and warn off other fish of the same species.
- There are over 120 species in the *Chaetodontidae* family.
- Some species, like Rainford's butterflyfish, are found in the GBRMP and Coral Sea, whilst others, such as the Banded butterflyfish, are found in the Caribbean.

In addition (and also within the *Chaetodontidae* family, but in different genera) are three other butterflyfish worthy of mention and illustrated on this page! These are:

Longfin bannerfish *(Heniochus acuminatus)* – IUCN Redlist = LC
Beaked coralfish *(Chelmon rostratus)* – IUCN Redlist = LC
Pyramid butterflyfish *(Hemitaurichthys polylepis)* – IUCN Redlist = LC

These species are sometimes seen in **schools** and are **planktivorous** in some settings!
This means that they feed on tiny **plankton** in the water column.
They share many of the **characteristics** of the *Chaetodon* genus, but not all, which is why they are classified differently!

FOR THE BUDDING SCIENTISTS AND CITIZEN SCIENTISTS, HERE IS A BIT OF FUN TO HELP YOU TO UNDERSTAND HOW LIVING THINGS ARE CLASSIFIED!

"***Kids Prefer Cheese Over Fried Green Spinach***"!

This is a popular mnemonic device for remembering the biological taxonomy: **K**ingdom, **P**hylum, **C**lass, **O**rder, **F**amily, **G**enus, **S**pecies.

This phrase helps students memorise the scientific classification system.

Far eastern curlew

Numenius madagascariensis

- Far eastern curlews, recognised by their "cuuuurlew" call, are the largest shorebirds in Australia, reaching heights of up to 60cm.
- The far eastern curlew begins its journey from its breeding grounds in Russia and northeastern China, travelling over 10,000km to reach warmer climates in southern feeding areas. On passage, they are commonly seen in Japan, Korea and Borneo.
- Even though they are only six to eight weeks old, the baby chicks know how to find their way on their very first **migration**! They have a special built-in sense of direction and distance.
- They rely on the **coastal regions** of the Great Barrier Reef Marine Park (GBRMP) as vital non-breeding grounds. In the GBRMP, they are primarily found in sheltered **intertidal** zones where they fuel up for their massive annual **migration** to the Northern Hemisphere. This high-energy diet is essential; they can lose 40–70% of their body weight during their nonstop flight from breeding grounds. However, the development and erosion of these mudflats along their **migratory** paths pose a significant threat to their survival.
- They feature an impressive, **distinctive** curved beak averaging 185mm in length, which is used to **probe** the mud and dig up shrimp, prawns and crabs, their main food source in Australia. Eastern curlews also eat **molluscs** and worms. During their **migration**, they even eat berries.
- Far eastern curlews are a bulky, dark-streaked brown wader with a long neck and legs.
- When flying, the barred flight feathers are visible, lighter under the wings and dark above. It has mottled brown and tan **plumage**. The average size is 63cm, and the average weight is 900 grams.
- Their wing beats are slow and deliberate, unlike the rapid beats of the Whimbrel.
- In Australia, the Eastern curlew is classified as critically endangered, with their **global population** declining by approximately 80% over the past three decades.
- Eastern Curlews are quite shy and easily disturbed, so it's important to give them space and keep pets on leashes when you're near their feeding or **roosting** areas, and avoid using drones nearby. They are **wary** birds, quick to take flight.

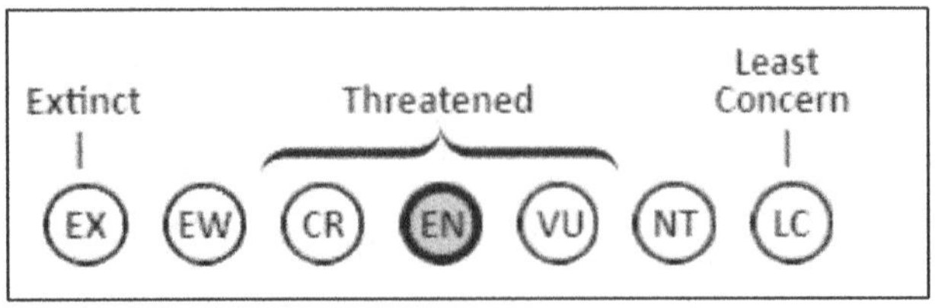

[1]

NCA List = CR – Critically endangered
EPBC List = CE – Critically endangered

Australian fairy tern

Sternula nereis exsul

- There are three **subspecies** of fairy tern: the Australian fairy tern, the New Zealand fairy tern and the New Caledonian fairy tern. It is the New Caledonian fairy tern which is found in the Coral Sea region, with breeding populations identified on East Diamond Islet near the Lihou Reef National Nature Reserve within the GBRMP region.
- These birds, which are **distinct** from southern Australian **populations**, were confirmed breeding in this area in 2006.
- While the more common Australian **subspecies** (*S. n. nereis*) is typically found along Australia's southern and western coastlines, the New Caledonian **subspecies** is a **distinct population** that has been documented making significant **trans-oceanic** movements to reach the Great Barrier Reef.
- These terns prefer remote coral cays and sandbanks for nesting, often **foraging** in nearby shallow, turquoise waters.
- They are small, white seabirds with pale grey wings and a black cap (in breeding plumage), with a noticeably yellow-orange bill.
- The fairy tern is approximately 22–27cm in length, 70g in weight and has a **wingspan** of 44–53cm
- The fairy tern is much smaller than almost all other Australian terns.
- The fairy tern mainly feeds on fish, which it catches by hovering, beating its wings **rapidly**, 3 to 10m above the surface of the sea before diving, beak first, into the water. It also feeds on **crustaceans** and **molluscs**.
- Like other **subspecies**, they face threats from human disturbance at breeding sites, extreme weather events like cyclones that wash over low-lying **cays**, and **predation**.

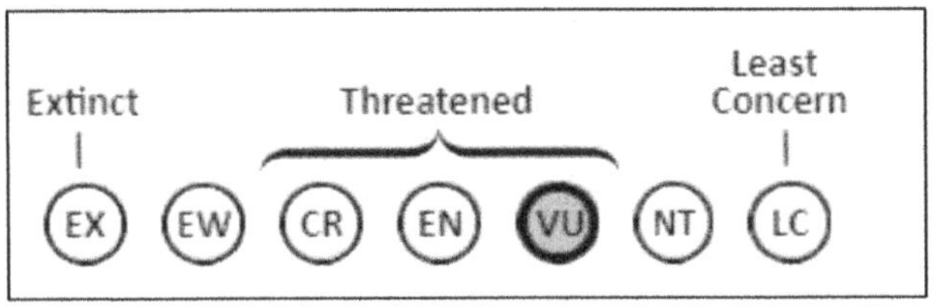

[1]

NCA List = EN = Endangered
EPBC List = E = Endangered

10,000 Birds:

Have a look at this list, it is so interesting, and has regular updates, as well as emails, focusing on different species of birds all around the world! Why not subscribe to their email list to learn more about all sorts of birds in all sorts of environments?!

- Here is the website address: (https://www.10000birds.com/about)

Palm cockatoo, goliath cockatoo

Probosciger aterrimus macgillvrayi

- The palm cockatoo (also known as the goliath cockatoo) is a large, smoky-grey parrot native to New Guinea and Cape York Peninsula, and therefore occurs in **coastal** areas of the Great Barrier Reef Marine Park. On Cape York, they are **revered** as "spirit birds" by several language groups who consider it bad luck to harm them.
- **Genetic** studies suggest they are one of the earliest offshoots of the cockatoo family, dating back millions of years. The Australian **population** is listed as **vulnerable** under the EPBC Act, with fewer than 2,000 individuals estimated to remain in the wild.
- They feature a **prominent**, punk-like crest, bright red cheek patches that blush when excited. They possess one of the largest beaks of any parrot (second only to the hyacinth macaw), capable of cracking nuts that humans would need a hatchet to open.
- Renowned as the only non-human species known to create musical instruments, males craft drumsticks from sticks and tap rhythmic beats on hollow trees to attract mates, sometimes for over 10 minutes; they can also use large seed pods. These birds are celebrated as the "rockstars" of the rainforest due to their extraordinary ability to craft and play musical instruments. Each male develops a unique **rhythmic** pattern, like a musical signature, that he uses to attract mates and mark territory.
- Their jaws do not meet, allowing them to use their upper beak and tongue to hold and break open incredibly tough nuts. They are the only parrot that can break open some of the larger seeds found in their **habitat**.
- The bare, red skin on their cheeks turns bright scarlet when the bird is alarmed or excited.
- They only lay one egg every two years, which is one of the lowest breeding success rates for any parrot species.
- With a **wingspan** of over 1.1m, they are Australia's largest parrot.
- They can live for many **decades**, with some individuals in **captivity** reportedly living for over 80 years.
- Considered a **primitive** species, they are the sole member of the genus *Probosciger*.
- They have one of the lowest breeding rates of any parrot, laying only a single egg every two years; on average, a pair may only successfully raise one chick per decade. They form strong, **monogamous** pair bonds and often occupy the same breeding territory year-round.
- They inhabit **tropical** rainforests and woodlands and rely on mature trees for nesting.
- They often place sticks into tree hollows to create a sturdy platform for their egg.
- They emit a wide variety of sounds, including a sharp screech when alarmed and various whistles.
- Their population is declining due to **habitat destruction** from mining and development, as well as their very low **reproductive** rate.

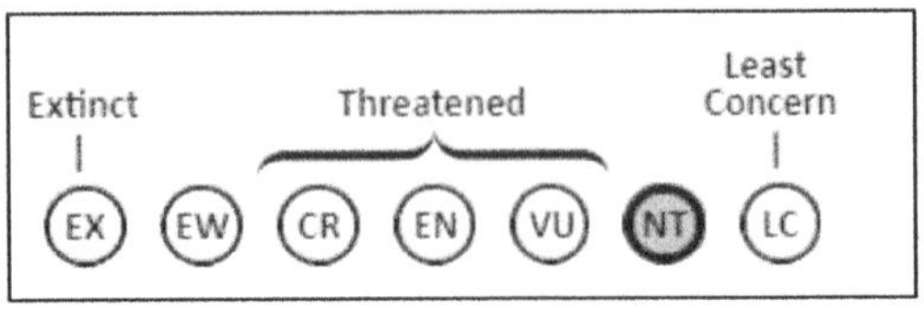

NCA List = EN = Endangered
EPBC List = V = Vulnerable

Parrotfish sp.

Family: Labridae
Subfamily: Scarinae
Genera:

- *Bolbometopon*
- *Calotomus*
- *Chlorurus* — *Illustrated: Chlorurus frontalis* **(Reefcrest parrotfish)**
- *Cryptotomus*
- *Hipposcarus*
- *Leptoscarus*
- *Nicholsina*
- *Scarus* — *Illustrated: Scarus chameleon* **(Chameleon parrotfish)**
- *Sparisoma*

- Parrotfish are a **clade** (a **fundamental** grouping in **evolutionary biology** consisting of a single common **ancestor** and *all* of its **descendants**) of fish placed in the **subfamily** *Scarinae* of the wrasse family (*Labridae*). Formerly treated as their own family (*Scaridae*), **genetic** studies found them to be deeply nested within the wrasses, and they are now treated as a **subfamily** (*Scarinae*). With roughly 95 species, the group's largest species richness is in the Indo-Pacific. They are found in coral reefs, rocky coasts, and seagrass beds, and can play a significant role in **bioerosion** because they eat away at the corals.
- Parrotfish are essential, **vibrant** coral reef cleaners that create white sand beaches, with one fish producing up to 800 pounds of sand **annually** by eating **algae**-covered coral. Using a beak-like, incredibly strong, regenerating set of teeth, they grind up coral.
- Their teeth are fused into a "beak," harder than a penny and capable of withstanding 530 tons of pressure, allowing them to eat stony coral. Beside their beak, they have a second set of teeth in their throat for grinding, with around 1,000 tiny teeth in total arranged in 15 rows.
- At night, some species create a bubble-like mucus **cocoon** to protect themselves from **predators** by masking their scent.
- Parrotfish are **hermaphrodites**; they can start as females and change into males as they mature, often changing colour to a brighter, more **vibrant** shade in the process.
- They eat **algae** off coral, which prevents the coral from being smothered and allows the reef to thrive.
- They use their **pectoral** fins to swim, a method known as 'rowing'.
- They **ingest** coral to reach the **algae** inside, which their bodies then turn into sand and **excrete**.

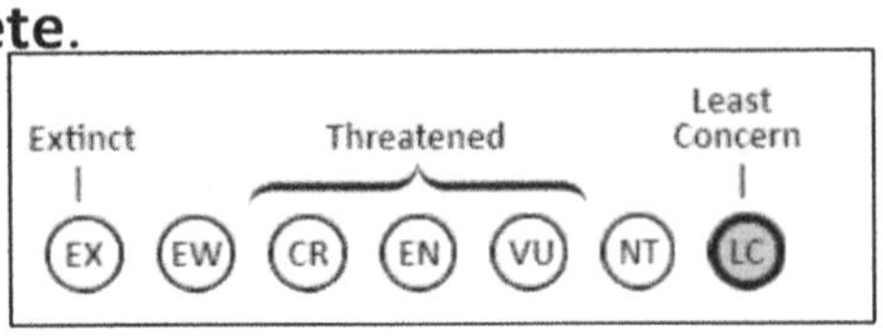

[1]

Whale shark

Rhinocodon typus

- While Ningaloo Reef in Western Australia is famous for consistent, large-scale **aggregations**, the Great Barrier Reef is increasingly recognised as a spot for sightings.
- Whale sharks visit the Great Barrier Reef Marine Park, particularly during summer (November to December), around coral and fish **spawning** seasons. Sightings frequently occur in the northern sections, including the Ribbon Reefs and off Townsville. While **elusive**, sightings are rising, with a known hotspot for **juvenile** males identified in Wreck Bay in the far north.
- Despite the name and size, they are fish, not **mammals**. They have **gills**, not lungs.
- Whale sharks are **ovoviviparous**, which means females produce eggs that **hatch** inside them. They give birth to live young and can produce around 300 babies at a time. These sharks do not start **reproducing** until they are around 30 years old. Female whale sharks are usually larger than males, as is true for most shark **species**.
- Each whale shark has a **unique** pattern of spots and stripes, allowing researchers to identify individuals, similar to a human fingerprint.
- To protect their eyes, whale sharks have small teeth (**dermal denticles**) covering their eyeballs. They can also retract their eyes into their heads.
- They can process over 6,000 litres of water per hour through their **gills** to eat **plankton**, **krill**, and small fish, using their roughly 3,000 tiny teeth for filter-feeding rather than biting.
- They typically swim at a slow 5k/hr but can dive down to 1,000m.
- Their **skeleton** is made of **cartilage** rather than bone. All sharks, skates and rays are **cartilaginous**. Like the rings of a tree, scientists can determine a whale shark's age by counting the growth rings in their **vertebrae**.
- Despite their 1.5m wide mouth, their throats are only about the size of an orange.
- The IUCN has listed the whale shark as **endangered**. Rising temperatures, **pollution** and ocean **acid**ification threaten whale sharks, which are also frequently caught in fishing nets and killed by ships' propellers, so human threats are also a factor in their population decline.
- Small yellow fish, likely **juvenile trevally** (*Gnathanodon speciosus*), are shown riding in or near the shark's mouth, which allows them to **conserve** energy by riding the bow wave and gain protection from predators.

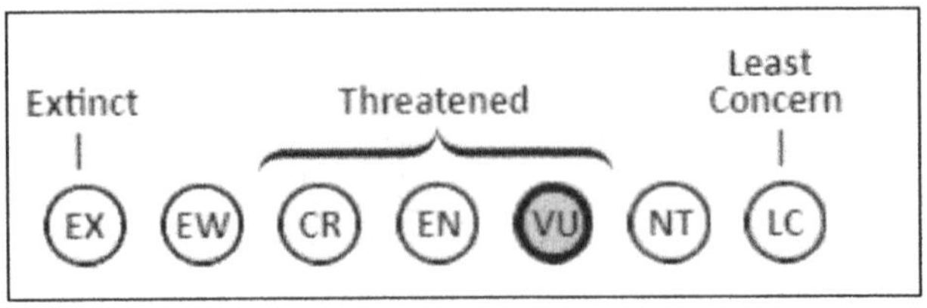

[1]

EPBC List = E = Endangered
NCA = Not listed

Surgeonfish, tangs, sawtails and unicornfishes

Family: Acanthuridae

- Surgeonfish, tangs, and unicornfish belong to the family *Acanthuridae*, a group of roughly 86 species of **tropical marine** fish known for their **vibrant** colours, **crucial** role in reef maintenance, and **specialised self-defence mechanisms**. They are found worldwide in **tropical** seas, usually around coral reefs.
- The family name comes from Ancient Greek ἄκανθα (*ákantha*), meaning "spine", and οὐρά (*ourá*), meaning "tail". Surgeonfish are named for the extremely sharp, **retractable** spine(s) located on either side of their tail base (**caudal peduncle**). These spines resemble a surgeon's scalpel. When threatened, they flick their tails to slash attackers, capable of causing deep wounds to **predators**—and careless handlers. While not generally **aggressive** toward humans, they can inflict deep, painful **lacerations** if handled.
- Many species have bright colours or contrasting patches at the base of their tail to warn **predators** of the **venomous**, razor-sharp weapon. Most surgeonfish/tangs have one spine on each side, but unicornfish (*Naso* genus) typically have two fixed blades on each side.
- Surgeonfishes sometimes feed as **solitary individuals**, but they often travel and feed in **schools**, which may be a **mechanism** for overwhelming the highly **aggressive defence** responses of small **territorial** damselfishes that **vigorously** guard small patches of **algae** on coral reefs.
- Most **species** are fairly small, with a maximum length of 15–40cm, but some in the genus *Acanthurus*, some in the genus *Prionurus*, and most species in the genus *Naso* may grow larger; the whitemargin unicornfish (*Naso annulatus*) is the largest species in the family, reaching a length up to 1m.
- These fish are the only known hosts of the **symbiotic bacteria** of the genus *Epulopiscium*. These **bacteria** affect the digestion of surgeonfish, enabling them to break down the **algae** in their diet.

Illustrated on the facing page are:

- *Naso lituratus* **(Pacific orange-spine unicornfish)**
- *Naso unicornis* **(Bluespine unicornfish)**
- *Naso annulatus* **(Whitemargin unicornfish)**
- *Naso brevirostris juv.* **(Spotted unicornfish)**
- *Naso hexacanthus* **(Sleek unicornfish)**

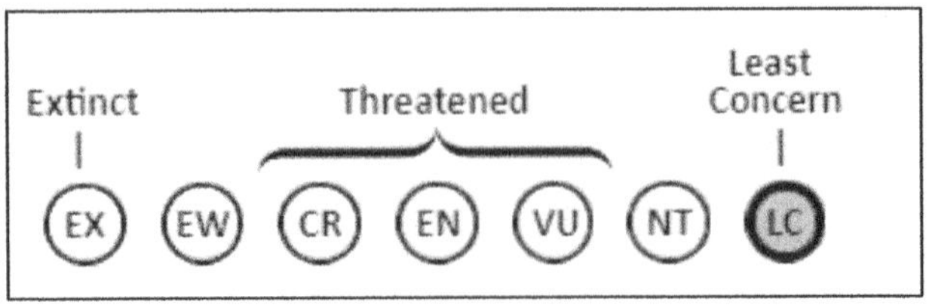

[1]

ALL SPECIES ON IUCN REDLIST ARE CLASSIFIED AS LEAST CONCERN

.... Continued on page 36.......

Surgeonfish, tangs, sawtails and unicornfishes (continued)

Family: Acanthuridae

- Some unicornfish develop a large, bony **protrusion** (horn) on their forehead. Not all unicornfish have horns. Some species develop them, others have a rounded, bump-like forehead, and some have no horn at all. The horn starts growing when the fish is about 12.5cm long. While the horn seems like a **defensive** tool, it is not used in defence. The fish still rely on their tail spines, and the exact purpose of the horn remains unknown, although it may be used in mating displays.
- Most surgeonfish and tangs are **herbivores**, feeding on **algae** and **marine** plants, which prevents **algae** from overgrowing and smothering corals. Some species have a specialised, tough, **gizzard**-like stomach that grinds down **algae**, sand, and coral **ingested** during feeding.
- Unlike many surgeonfish, many unicornfish (*Naso* species) are **diurnal planktivores**, feeding on **plankton** in the water column above the reef, though they still eat **macroalgae**.
- Surgeonfish can change colour based on their mood, stress levels, or time of day. Males often darken during breeding or when competing for **dominance**.
- Many species of acanthurids are long-lived, with some exceeding 30 years in the wild.
- They are **broadcast spawners**, often **aggregating** in huge numbers (sometimes thousands) to release eggs and **sperm** simultaneously into the current.
- Their larvae, known as "**acronurus**", are completely transparent and drift in the open ocean for a long time before settling on a reef and **maturing**.
- "Dory" the Tang: The famous character Dory from *Finding Nemo* and *Finding Dory* is a Palette Surgeonfish (also known as a Regal Blue Tang or Hippo Tang).
- "Tang" is a name usually applied to specific **genera**, such as *Zebrasoma* (like the **vibrant** Yellow Tang) and *Acanthurus* (like the Convict Tang). *Acanthurus* is from the Greek "akantha" (thorn) and "ourá" (tail) = thorned tail, in reference to the blade-like structures on the **caudal** fin base of surgeonfishes. The name "tang" comes from Old Norse and early Scandinavian roots. Historically, a "tang" was a sharp projecting point or prong on a tool, such as the metal spike extending into a knife handle.
- Bristletooths: *Ctenochaetus* tangs are specialised to eat **detritus**, **algae**, and film from rocks using brush-like teeth.
- They are a common food source in many Pacific Island coastal fisheries.
- Some surgeonfish, particularly those that eat certain types of **algae**, can accumulate toxins, causing **ciguatera** (fish poisoning) if eaten.

Illustrated on the facing page are:

Acanthurus triostegus **(Convict tang)**
Zebrasoma scopas **(Twotone tang)**
Zebrasoma veliferum **(Sailfin tang)**
Acanthurus nigrofuscus **(Brown tang**

SURGEONFISH

<u>*Illustrated on the this page are:*</u>

Acanthurus pyroferus **(Chocolate surgeonfish)**
Acanthurus dussumieri **(Eyestripe surgeonfish)**
Acanthurus guttatus **(Whitespotted surgeonfish)**
Acanthrus lineatus **(Striped surgeonfish)**
Acanthurus mata **(Elongate surgeonfish)**
Acanthurus olivaceus **(Orangeband surgeonfish)**

PACIFIC BLUE TANG AND SAWTAILS

Illustrated on the this page are:

Paracanthurus hepatus **(Regal blue tang), "Dory" in the film "Finding Nemo"**

Prionurus maculatus **(Yellowspotted sawtail)**

Victoria's riflebird

Ptiloris victoriae

- Victoria's Riflebird is one of Australia's most spectacular birds. As a member of the bird-of-paradise family, it is renowned for its intense **courtship** dances and breathtaking **plumage**. It is one of the only birds-of-paradise **endemic** to the Australian mainland.
- Whilst it is said to live exclusively in the Wet Tropics rainforests of northeastern Queensland, particularly around the Atherton Tablelands and Daintree Rainforest, Atlas of Living Australia records indicate reported and recorded sightings in **coastal** areas within Eastern Kulu Yalanjo Indigenous Protected Area sea country and Girringun Indigenous Protected Area sea country, which fall within The Great Barrier Reef Marine Park.
- Measuring only 23–25 cm, it is the smallest of all riflebird species.
- The male's plumage is not just black; it is a specialised velvet black that absorbs 99.95% of light, making them look like a "black hole" in the dense, dappled light of the rainforest. While the body is black, the throat and head boast intensely **iridescent** emerald-green and peacock-blue feathers that flash brilliantly during display.
- It takes young males 4–5 years to develop this full adult **plumage**, during which they resemble females.
- Males are **polygynous**, meaning they mate with as many females as possible. To attract them, they perform an intense, **mesmerising courtship** dance. A male will sit on a "dancing perch" (often a bare, dead stump), raise his wings into a high circular fan, and sway side-to-side, flicking his head from one side to another while producing a loud, raspy "yaars" call. The **courtship** often involves the male encircling the female and almost hitting her with his wings.
- The name "riflebird" comes from the 19th-century belief that their black plumage resembled the uniforms of the British Army Rifle Brigade. Another interpretation is that their call sounds like a rifle bullet flying through the air.
- Unlike many birds that pick insects from leaves, the Victoria's Riflebird behaves more like a woodpecker or treecreeper. They use their long, curved, needle-like beaks to **probe**, tear, and search for spiders and insects in the bark and rotting wood of trees.
- The female is olive-brown with a pale, barred underbelly. She is solely responsible for building the nest, incubating the eggs, and raising the young. Females often line their nests with shed snakeskin, a common habit among birds-of-paradise, which may **deter predators**.
- Despite their small size, Victoria's Riflebirds can live for a surprisingly long time. A wild male has been recorded living for more than 15 years.
- The species was discovered in the 1840s by John MacGillivray and was formally described by John Gould in 1850, who named it in honour of Queen Victoria.

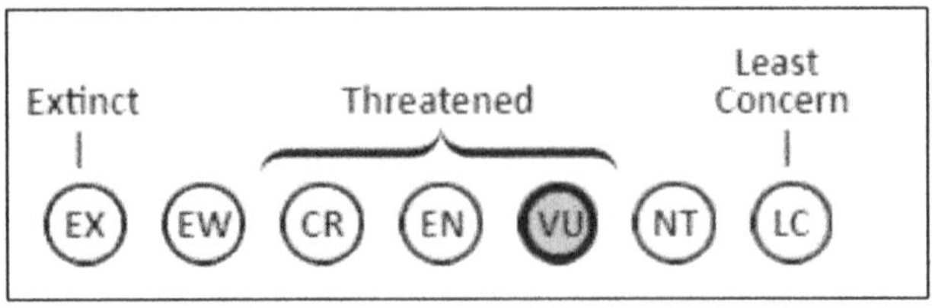

[1]

EPBC List = Not listed
NCA = Not listed

Capricorn silvereye

Zosterops lateralis chlorocephalus

- The Capricorn silvereye or Capricorn white-eye is a distinct subspecies of silvereye **endemic** to the **coral cays** of the Capricorn and Bunker Groups on the Great Barrier Reef. These **sedentary** birds are known for their bright olive-green heads, **unique** island **ecology**, and pair-bonding behaviour, sometimes considered a **distinct** species.
- They are the only bird species **endemic** to the Great Barrier Reef, restricted entirely to the wooded **coral cays**, particularly in the southern region.
- They are recognised by a bright olive-green head, back, and upper wings, which are more vibrant than those of many mainland silvereyes, along with the **characteristic**, **prominent** white ring of feathers around the eyes.
- Unlike the migratory Tasmanian silvereye populations, the Capricorn silvereye does not **migrate** and lives permanently on specific islands.
- Studies on Heron Island show these birds mate for life with the same partner.
- They form a rapidly **differentiating metapopulation**, with **genetic isolation** from other *Z. lateralis* subspecies, making them a significant subject for studying rapid **evolution**.
- Their diet consists of small insects, fruit, and **nectar**, often **foraged** in pairs or small family groups.
- They were first described by **ornithologists** Archibald Campbell and Samuel White in 1910 following an expedition to the Capricorn Group.
- They are known for a "subdued chorus of sweet warbling songs" at daybreak.
- They are efficient breeders, laying 2–4 pale blue eggs that hatch in roughly 11 days.
- These birds are often considered a key indicator species for the **ecological** health of the **coral cay ecosystems** they **inhabit**.

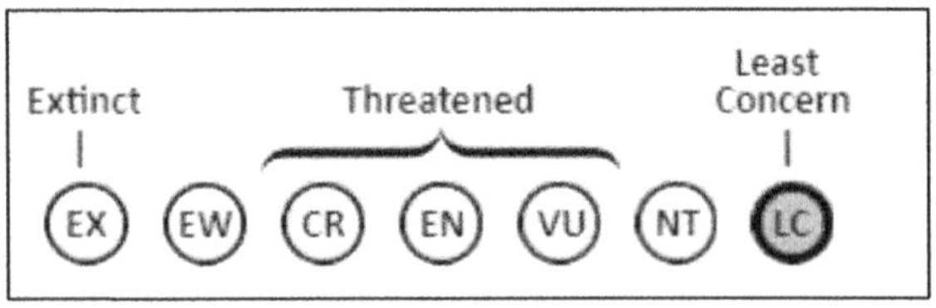

[1]

EPBC = Not listed
NCA List = Not specifically listed

Halifax fan palm

Livistona drudei

- The Halifax fan palm is a native palm species found in **coastal** areas adjacent to the Great Barrier Reef World Heritage Area, including Hinchinbrook Island National Park.
- Endemic to northeast Queensland, found in coastal lowlands between south of Innisfail, Hinchinbrook Island, Rollingstone, and Proserpine.
- Occupies melaleuca swamp-forests and the fringes of gallery or tropical rainforests bordering eucalypt forest. It is frequently found along streams in coastal plains.
- It is listed among the rare or threatened plants within the Great Barrier Reef World Heritage Area/Catchment areas.
- As of May 2024, it is considered Vulnerable under Queensland's Nature Conservation Act (NCA) but is listed as Endangered under the IUCN Red List.
- A tall, slender palm that can grow to 28 m in height. It is often distinguished by its large root mass, grey-green leaves, and yellow-orange flowers.
- The palm is known to **inhabit** areas often **inundated**, and it is frequently found in freshwater-influenced areas rather than on the reef islands themselves.

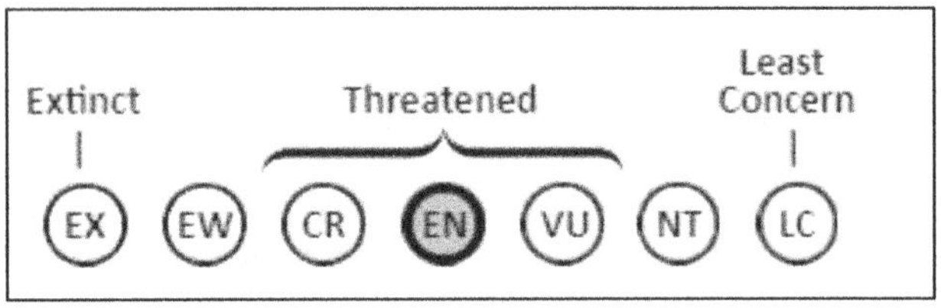

[1]

EPBC List = Not listed
NCA = V = Vulnerable

Green sawfish

Pristis zijsron

- The green sawfish, also known as the longcomb sawfish, is a remarkable, critically endangered species of fish often mistaken for a shark. It belongs to the family Pristidae and is one of the largest rays in the world.
- The green sawfish is potentially the largest of all sawfish species, with recorded lengths up to 7.3 meters.
- This species holds the record for the longest **rostrum** (the "saw") of any modern sawfish species. The saw is narrow and studded with 24–28 pairs of unevenly spaced teeth.
- Despite their shark-like appearance, green sawfish are actually a type of specialised ray. Their **gill** openings are located on their underside, unlike sharks, which have them on the sides.
- The **rostrum** is covered in thousands of tiny sensory organs called **ampullae of Lorenzini**. These allow the sawfish to detect the weak electric fields produced by their **prey** (fish and **crustaceans**) in **murky** water.
- The saw is not just for smelling. They use it to dig in the sand for food and to thrash around to stun schools of fish before eating them.
- Green sawfish are long-lived, with a **lifespan** that can exceed 50 years. They are slow to **mature**, with females estimated to reach **maturity** around 9 years of age.
- They are **ovoviviparous**, meaning they give birth to live young rather than laying eggs. Females usually have litters of about 12 pups, which are born with their "saws" already developed (though the saw is flexible at birth to protect the mother).
- They are found in **tropical** and **subtropical** waters, often inhabiting coastal areas, **estuaries**, and, surprisingly, **brackish** and freshwater systems.
- The green sawfish is listed as critically endangered on the IUCN Red List. Their numbers have declined significantly (estimated over 80%) due to accidental capture in fishing nets, **habitat degradation**, and demand for their saws in traditional medicine and the curio trade.
- In some areas, such as New South Wales, Australia, the green sawfish has not been seen in the wild since 1972 and is considered to be locally extinct.

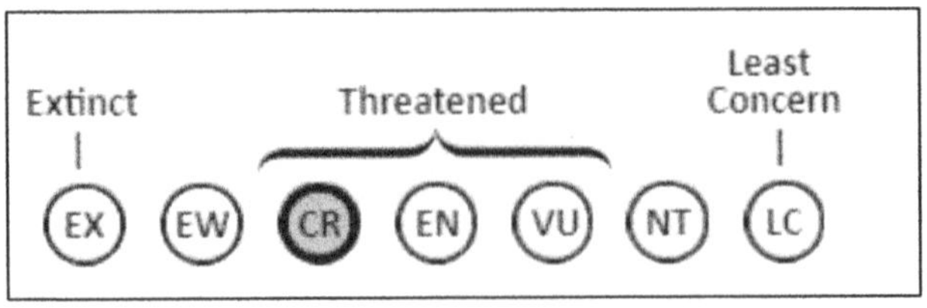

[1]

NCA List = Not listed
EPBC List = V = Vulnerable

Porkchop coral, tombstone coral

Pavona duerdeni

- *Pavona duerdeni* is a slow-growing stony coral that forms massive, large colonies with parallel ridges or hillocks in shallow Pacific reefs. Key facts include its ability to withstand high-flow, **murky**, and high-tidal environments by eating suspended **organic** particles. These corals create extremely dense, huge **skeletal** structures.
- These coral colonies are known for having very dense **skeletons**, which indicates a slow, steady growth rate over many years. This species forms large, cream-coloured, solid colonies made up of lobed or ridged structures.
- Their massive, thick colonies form ridges or large hillocks, sometimes earning them the nickname "tombstone coral" for their appearance. In addition to tombstone, they are often referred to as porkchop coral because of the unique shape of their lobed or ridged structures.
- They often thrive in **murky**, high-flow environments between islands where they feed on **organic** particles in the water column.
- They are more commonly found in Hawaii, in shallow waters throughout the Indo-Pacific, including the Great Barrier Reef, and the Tropical Eastern Pacific. Studies of the region's nearshore **marine biodiversity** indicate that *Pavona duerdeni* and similar species are common in these tropical reef environments.
- Despite their massive size, they appear relatively smooth up close due to tiny, densely packed **corallites**. *Pavona duerdeni* is recognised by its distinctive, intricate star-shaped (**thamnasteroid**) **corallites**.
- These corals are a key part of the reef structure and are often studied for their ability to thrive in challenging, **turbulent** environments.

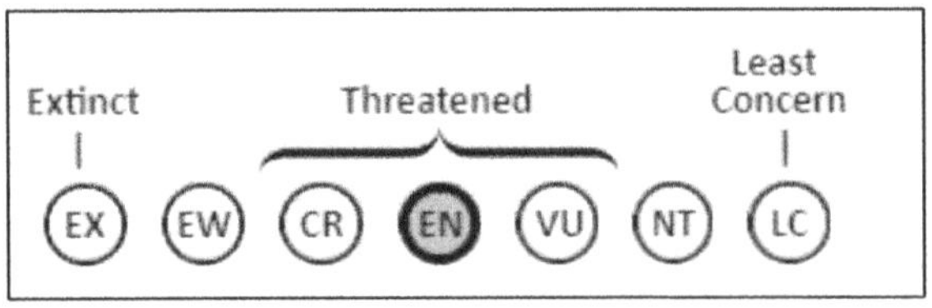

[1]

EPBC List = Protected
NCA List = Not listed

Illustrations:
Top – Colony formation
Bottom – Pattern of corallites

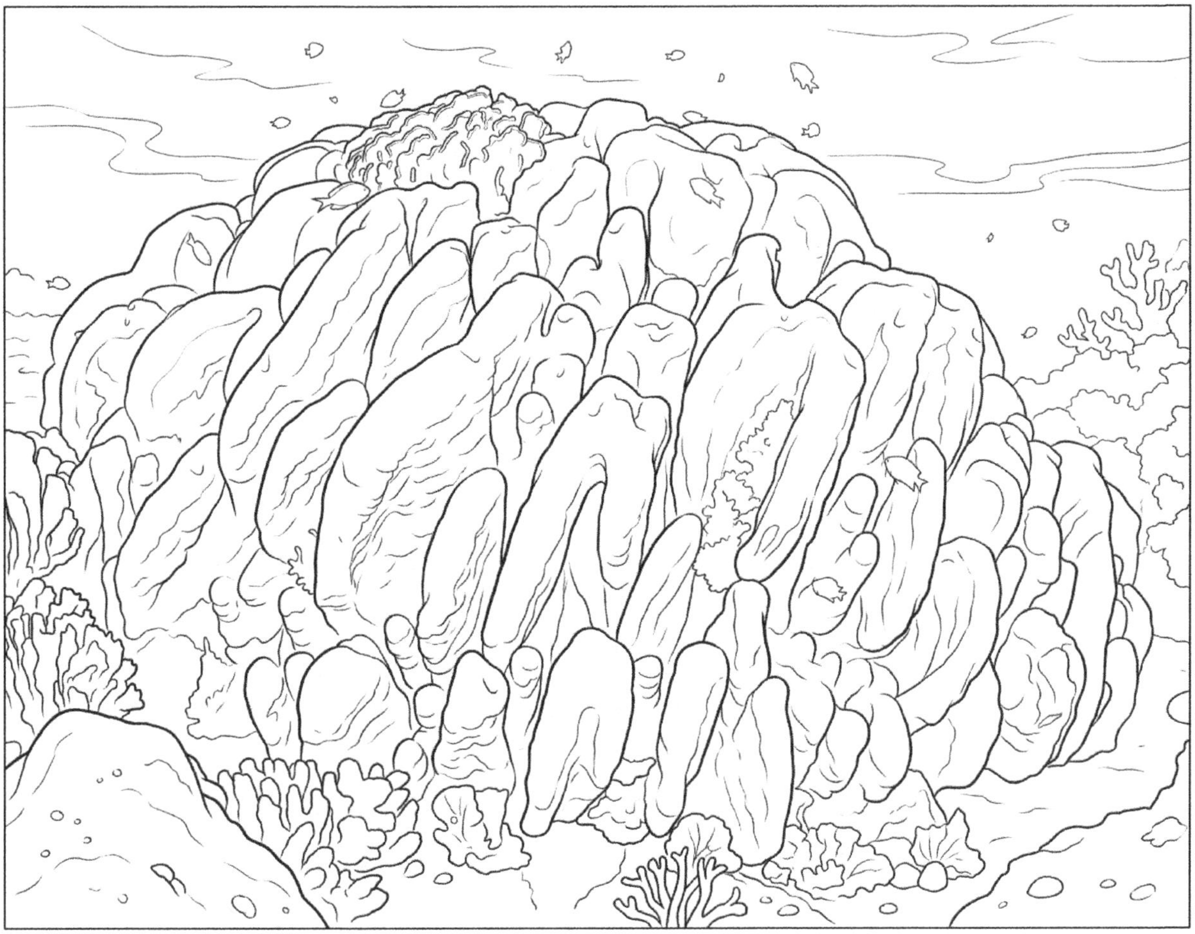

Humphead wrasse, Humphead Māori wrasse

Petrogale xanthopus

- The Humphead Māori wrasse is a massive, long-lived reef fish, growing over 2m long and living over 30 years. Known for a prominent forehead hump that grows with age, they are **protogynous hermaphrodites** (females can change into males). These friendly, curious fish are **crucial** for reef health because they eat **toxic** crown-of-thorns starfish.
- The largest member of the wrasse family, they can weigh over 180kg.
- They are born female and some transition to males at around 9 years old. The hump on their forehead becomes more prominent as they grow into males.
- Their name comes from the intricate, bright blue/green, geometric patterns on their faces, which look similar to traditional Māori facial tattoos.
- They are known to be curious and comfortable around divers, often approaching them.
- They are one of the few **predators** of the **poisonous** crown-of-thorns starfish and other **toxic** animals like sea hares and boxfish.
- They have strong, fused teeth, similar to a parrotfish, and secondary teeth in their throat for crunching hard shells like **crustaceans** and **molluscs**.
- Despite living long, they have a very slow breeding rate, becoming **mature** at 4-6 years.
- They are known to sleep in caves or under coral ledges at night.
- Due to high demand in the live fish trade and **habitat** loss, they are listed as Endangered on the IUCN Redlist.
- They are fully protected in Australian waters.

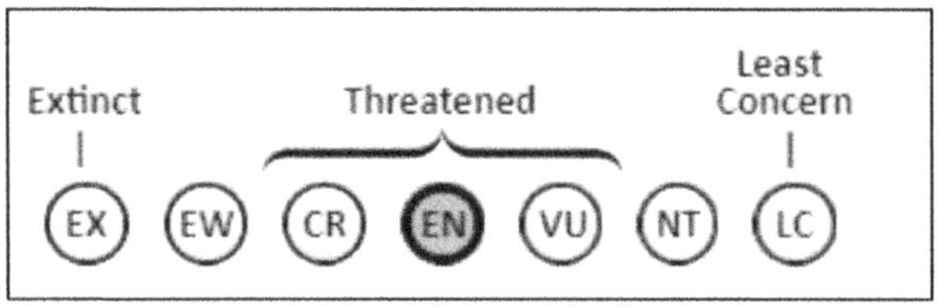

[1]

EPBC List = Not listed

NCA = Not listed

Reef manta ray

Mobula alfredi

- The Reef Manta Ray is the second-largest species of ray in the world and is renowned for its intelligence, **curiosity**, and **unique** feeding habits. The giant oceanic manta can reach a wingspan of 7-9m! Reef manta rays are slightly smaller, typically spanning 3-5.5m.
- Manta rays possess the largest brains of any fish species. They exhibit high levels of intelligence, long-term memory, and problem-solving abilities, similar to dolphins and primates.
- Similar to a human fingerprint, reef manta rays have unique spot patterns on their ventral (under) surface. Scientists use these markings to identify individuals and track their movements.
- They are filter feeders that consume **krill** and **zooplankton**. They often use a "chain feeding" technique, swimming in tight spirals to create a **vortex** that concentrates their **prey**.
- They are known for being very **curious** creatures and will often voluntarily interact with divers. They can also dive to depths of 300-400m, keeping their brains warm with a specialised counter-current heat-exchange system.
- Male reef manta rays bite the tip of the female's left pectoral fin during mating, leaving permanent scars. Over 95% of these mating scars are found on the left side, showing a strong "left-handedness" in their mating behaviour.
- Unlike their close cousins, the devil rays, reef manta rays do not have a **venomous** sting or spine on their tail.
- They have one of the lowest **reproduction** rates of any fish, often only giving birth to one pup every two to four years. They can live for around 40 to 50 years.
- They are known to breach or leap out of the water, hitting the surface with a loud thud. This behaviour is believed to be used to dislodge **parasites** or as a form of communication.
- *Mobula alfredi* was named after Queen Victoria's son, Prince Alfred, who survived an assassination attempt in Sydney in 1868.
- Unlike the Giant Oceanic Manta, the reef manta has a blurred, V-shaped white marking on its back and spots between its **gill** slits, making it distinguishable from its larger cousin.

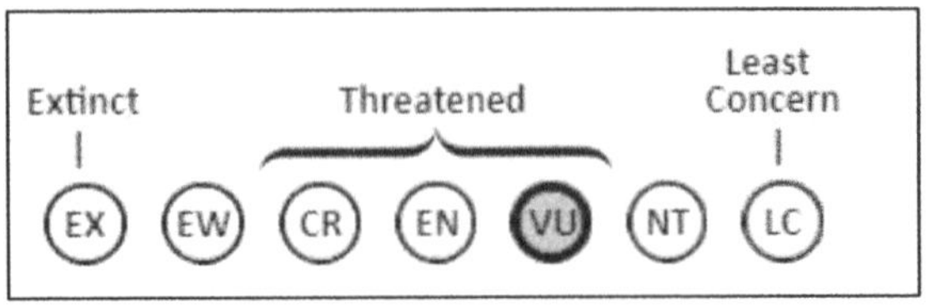

[1]

EPBC List = Migratory
NCA = Not listed

Scroll coral, cabbage coral

Montipora foliosa

- *Montipora foliosa*, often called Scroll Coral or Cabbage Coral, is a widely recognised small plating stony coral prized for its dramatic, leafy structures and rapid growth. It is a popular, relatively hardy choice for aquarium hobbyists looking to add structure to their tanks.
- This species forms thin, **laminar** (leaf-like) plates that often grow into complex, overlapping tiers, **whorls**, or vase-like shapes.
- Unlike other plating species like *Montipora capricornis*, *Montipora foliosa* is **characterised** by **conspicuous**, **radiating** ridges on its surface, which run **perpendicular** to the colony **margin**.
- *Montipora foliosa* is known to be a fast grower, with the ability to form large colonies, sometimes over one metre high in the wild.
- It thrives in high-light and moderate to high-flow environments. It is often found in shallow, protected reef slopes and lagoons at depths of 2–20m.
- While it is a common species, research suggests *Montipora foliosa* is highly susceptible to bleaching with low resistance, making it a target for monitoring in climate-impacted areas.
- It is found throughout the Indo-Pacific region, including the Great Barrier Reef, Indonesia, and the Maldives.
- While commonly seen as green, brown, or tan, this coral can exhibit vibrant colours such as cream, pink, or purple, especially in shallow, high-light areas.
- It is less **palatable** to many **corallivores** than other corals, but it is vulnerable to the "Montipora-eating **nudibranchs**," **parasites** that can cause rapid tissue damage.

  In the aquarium trade, *Montipora foliosa* is frequently confused with or sold under the name *Montipora capricornis*. While they have similar, overlapping growth forms, *Montipora foliosa* can be distinguished by the prominent surface ridges, whereas *Montipora capricornis* has a smoother surface.

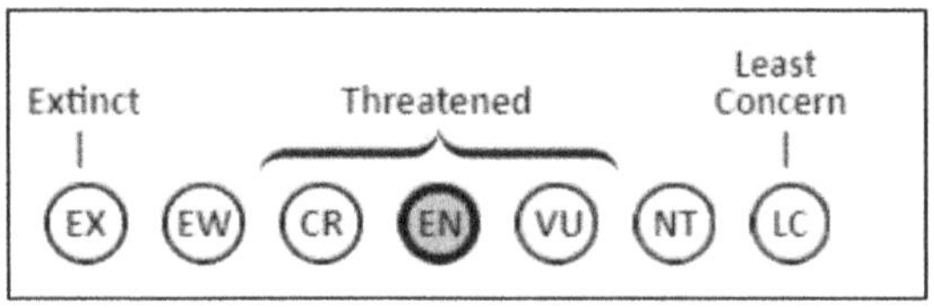

[1]

EPBC List = Not listed
NCA = Not listed

Illustrations:
Top – Colony formation
Bottom – Pattern of corallites

Sapphire damselfish, peacock damselfish, blue damselfish
Pomacentrus pavo

- *Pomacentrus pavo* is a **vibrant**, small Indo-Pacific reef fish known for its **iridescent** blue-green colour, dark "ear" spot, and striking yellow-tipped tail. Growing to only about 8.5–9cm, these fish **inhabit** sandy **lagoons**, rubble, and coral heads.
- They possess an intense, **iridescent** blue-green body that can lighten in shade depending on the time of year. During mating rituals, males can further **distinguish** themselves by darkening in colour and displaying white blotches.
- A key identifying feature is a distinct, small dark or black spot located behind the **gill** cover.
- Despite their small size and "jewel-like" appearance, they are extremely **territorial** and **aggressive**, known to chase or **harass** much larger fish.
- *Pomacentrus pavo* are egg-layers (**oviparous**) that form distinct pairs to **spawn** at dawn. The male meticulously cleans a site for a nest, guards the **demersal** eggs, and aerates them until they hatch.
- They are frequently found in shallow waters, generally 1–12m, around man-made structures like jetties, as well as in sandy rubble areas.
- They are **omnivorous,** feeding on **zooplankton**, fish eggs, and **filamentous algae**.

Neon damselfish
Pomacentrus coelestis

- *Pomacentrus coelestis* is a **vibrant**, small reef fish found in the Indo-Pacific, notable for its electric blue body and bright yellow tail. Despite being common in the aquarium trade, these **territorial** fish are known for their incredibly short lifespans.
- The species appears bright i**ridescent** blue in sunlight, often with a brilliant yellow tail and belly, allowing them to blend into **vibrant** reef environments.
- Studies on the Great Barrier Reef indicate these fish often live for only 127–160 days, contrary to many damselfish that live for several years.
- Due to their short, fast life cycle, they are rapid **colonisers** and can quickly **dominate** reef areas.
- Although small, growing up to 9–10cm, they are intensely **aggressive**, often chasing or **harassing** much larger fish in their territory.
- **Juveniles** are known for schooling amongst soft corals, while adults prefer to live in smaller groups close to rubble beds.
- During breeding, they form distinct pairs; males guard and aerate the eggs, which stick to the **substrate** (**demersal**).
- They primarily feed on **zooplankton** and sometimes **graze** on **benthic algae**.

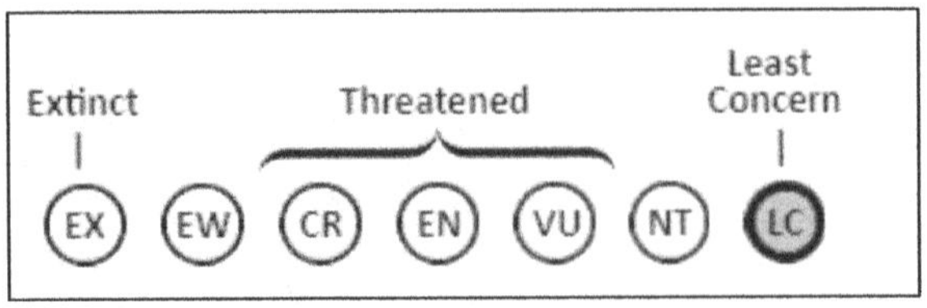

[1]

Clark's Anemonefish, Yellowtail Clownfish
Amphiprion clarkii

- While many clownfish are specialists, *Amphiprion clarkii,* is famous for its lack of **host specificity**. It is the only anemonefish that lives in association with all ten species of sea anemones known to host clownfish.
- Clark's Anemonefish displays the widest colour variation of any clownfish species, depending on geography, age, and **host** anemone. While they often appear black **dorsally** with orange ventral areas, they can range from entirely orange to nearly completely black. Their tail is generally a lighter colour, ranging from yellow to white.
- This species is known for producing audible sounds to communicate, especially when defending their territory or within their group. They make clicking and popping noises, which are easily heard by humans.
- Like all anemonefish, *Amphiprion clarkii* are **sequential hermaphrodites** and **protandrou**s, meaning they are born male and can change into females.
- They live in small groups consisting of a large, **dominant** female, a smaller, sexually functional male, and several smaller, nonfunctional **juveniles**.
- If the **dominant** female dies, the male will change sex to become the new female, and the largest **juvenile** will become the new functional male.
- While many reef fish have short lifespans, Clark's Anemonefish can live a long time. In the wild, they have been estimated to live up to 13 years, with some studies suggesting they may live even longer.
- Rarely, *Amphiprion clarkii* individuals develop a **unique mutation** where a small, pale blue or white crescent appears on the upper part of their iris, known in the aquarium trade as "Pearl Eyes".
- After a female lays eggs on a flat spot near the anemone, the male takes over the role of caretaker. He spends days fanning the eggs with his fins to ensure they get oxygen, and he will eat any dead or infertile eggs to protect the rest of the clutch from **fungus**.
- *Amphiprion clarkii* has the largest distribution of any anemonefish, found from the Persian Gulf across to Western Australia, and as far north as Japan.

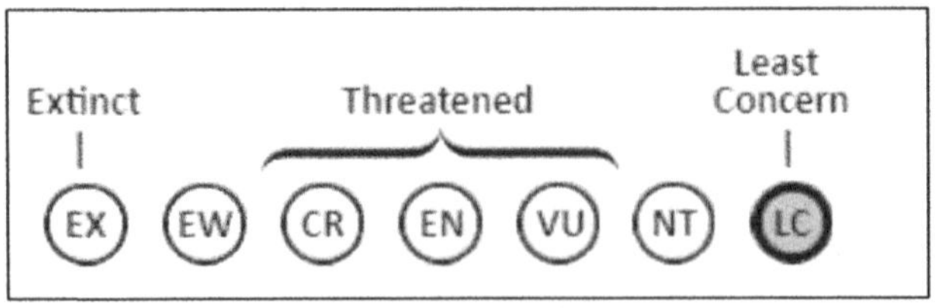

[1]

Loki whip goby

Bryaninops loki

- The Loki whip-goby is a tiny, highly specialised fish found in the Indo-Pacific. It is named after the Norse god of mischief because of its elusive nature and ability to disappear against its host corals.
- These fish can change colour over a few weeks to perfectly match the specific sea whip or **gorgonian** coral they live on, appearing in shades of grey, black, brown, cream, red, or yellow.
- The Loki whip goby is a **commensal species** that lives almost exclusively on **gorgonians** (sea whips) and black corals, often spending its entire life on one host.
- They are notoriously difficult for divers to spot, often only becoming visible to the naked eye after a photographer has taken a picture and looked closely at the screen.
- They are small, generally measuring less than 20mm in standard length, though they can reach up to 4.5cm.
- To live on swaying sea whips in high-current areas, they have evolved a cup-like structure for their **pelvic** fins to grip the coral tightly.
- They are often found on specific **hosts** such as *Junceela fragilis* in Indonesian waters.
- They occur in depths ranging from 6to 45m and are found across the Indo-West Pacific, from southern Japan to the Great Barrier Reef, Lord Howe Island, and even the Red Sea.
- They feed on **zooplankton** that drift past their **host** coral.
- The Loki whip goby is a prime example of **specialisation** and **adaptation** in coral reef **ecosystems**.

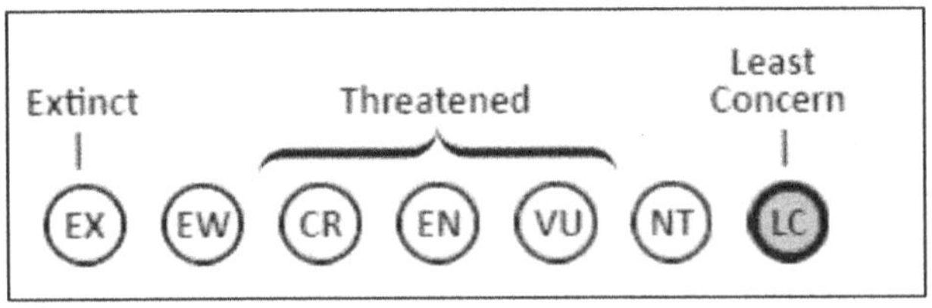

[1]

Clown Triggerfish

Balistoides conspicillum

- The Clown Triggerfish is one of the most visually **distinct** and sought-after marine fish in the world, renowned for its dramatic colouring and **unique** behaviours. Found in **tropical** Indo-Pacific reefs, these fish are highly intelligent and, as adults, tend to be **solitary** and **territorial**.
- The fish has a first **dorsal** fin with three spines, which they can lock into place to wedge themselves firmly into coral crevices and avoid being pulled out by **predators**.
- The first, large spine can only be unlocked by the fish itself when it engages a smaller second spine.
- To **intimidate rivals** or **predators**, they can produce a loud, grunting sound.
- Unlike many fish that use their tail for movement, triggerfish use a graceful, undulating motion of their second **dorsal** and **anal** fins to move.
- They are one of the few fish species capable of swimming backwards.
- They can create a **vortex** of sand to expose prey by spraying water from their mouth.
- They have a black belly with large, bold white spots, a yellow **reticulated** pattern on their back, and an orange-yellow snout, making them easily identifiable.
- Their eyes are set very high on their head and can move independently of each other, allowing them to scan for food and **predators simultaneously**.
- They have extremely strong teeth and jaws adapted to crush sea urchins, crabs, and hard-shelled **molluscs**.
- While **juveniles** are often found in small, secretive caves at depths beyond 20m, adults are typically **solitary** and prefer to live along steep, clear reef drop-offs at depths of 1–75m.
- Out of 40 species of triggerfish, the Clown Triggerfish is considered largely **inedible**, partly due to its diet.
- They are highly **territorial** and can be **aggressive** toward other fish, and even against their own kind.

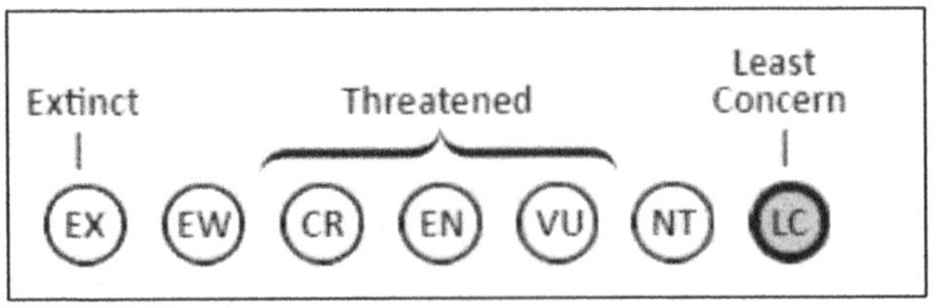

[1]

Laced moray, honeycomb moray, tesselate moray

Gymnothorax favagineus

- The laced moray, honeycomb moray, or tessellate moray, is a striking, large-bodied **marine** fish found in the Indo-Pacific. Known for its distinct black-and-white patterned skin, this eel is a master of **camouflage** and a **formidable** reef **predator**.
- The eel's name comes from its pattern of black, angular spots on a white or yellowish background, resembling a honeycomb. Interestingly, this pattern is not fixed—as they grow, the spots become smaller, and the overall look changes. Additionally, individuals living in clearer water tend to have fewer spots than those in **turbid** environments.
- The laced moray is among the largest of all moray eel species, capable of reaching lengths of up to 3m. They are second in size only to the giant moray in the Indo-Pacific region.
- Like other morays, they possess a second set of jaws called **pharyngeal** jaws inside its throat. When hunting, it uses its primary outer jaws to seize prey, and then the inner jaws shoot forward to snatch the prey and drag it down into the **oesophagus**, similar to the **mechanism** in the *Alien* movie franchise.
- When the laced moray catches a large prey item, such as an octopus or big fish, it has a special technique to consume it: it ties its own tail into a knot and slides this knot down its body until it hits the prey, creating **leverage** to tear the prey apart and swallow it.
- Despite being an **intimidating predator**, they often share their coral reef crevices with cleaner wrasses or shrimps. These small creatures clean the moray's teeth and remove **parasites** from its skin, getting a meal in return.
- They are frequently seen with their mouths wide open, which looks **aggressive** but is actually a breathing **mechanism**. The constant opening and closing allows them to pump water over their **gills**.
- Large specimens are known to accumulate high levels of a toxin called **ciguatoxin**, which is **accumulated** from eating **herbivorous** fish that consume **toxic** seaweed. Eating a large laced moray can cause severe food poisoning in humans called **ciguatera**, which can cause **intense neurological** and **gastrointestinal** issues.
- They are primarily active at night, leaving their hiding spots to hunt along the reef. During the day, they remain securely tucked away in holes and crevices.
- Instead of scales, the laced moray has smooth skin covered in a thick layer of yellow **mucous**, which helps it glide effortlessly through tight, sharp rocky reef areas.
- While generally shy and non-**aggressive** to divers, large, established adults can be **territorial** and **aggressive** if **provoked** or disturbed in the wild.
- Because they are popular in the **marine** trade due to their appearance, it is known that they are escape specialists in **captivity**, capable of moving lid covers to escape, especially when they are younger and smaller.

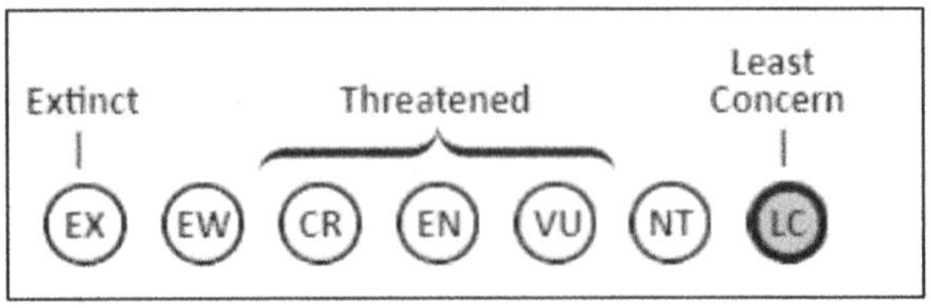

[1]

Spotted toby, honeycomb toby

Canthigaster janithinoptera

- The honeycomb toby, or spotted toby, is a small, fascinating species of sharpnose pufferfish found in Indo-Pacific reefs.
- They have a dark, brownish body covered in **numerous** small, pale bluish, greenish, or white spots that form a pattern resembling a honeycomb.
- They are known for having bright, contrasting colours, with males sometimes displaying more **vivid** spots and lines around the eye.
- While some pufferfish grow very large, the honeycomb toby is a dwarf species, with a maximum length of only about 8-9cm.
- Like other pufferfish, they have four teeth fused into a beak-like structure, which they use to feed on sponges, **algae**, **tunicates**, marine worms, and small corals.
- When threatened, they can puff up their bodies to nearly double their size, making them difficult for p**redators** to swallow. They also contain **tetrodotoxin**, a powerful **neurotoxin**, making them **poisonous** to eat.
- In the wild, they are very shy and often hide in crevices, holes in coral, or under overhangs. They are typically found alone or in pairs in **lagoons** and seaward reefs, usually at depths of 1-30m.
- They lack **pelvic** fins and swim by using their **pectoral**, **dorsal**, and **anal** fins, giving them a hovering, delicate, yet highly **manoeuvrable** swimming style.
- While they can be shy, they are known to have a "wonderful personality" in aquariums, though they are not considered safe for reef tanks because they may nip at corals and **ornamental invertebrates**.

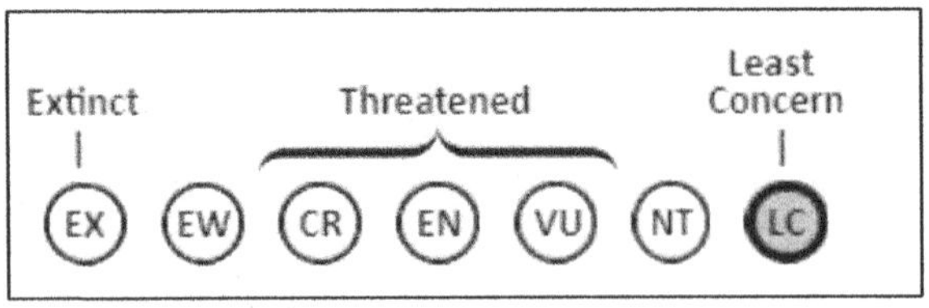

[1]

Triggerfish

Family: Balistidae

- Triggerfish (family *Balistidae*) are highly intelligent **marine** fish known for their **unique** "locking" **dorsal** spine, **aggressive** nest protection, and specialised hunting techniques, such as blowing water to uncover prey. They are surprisingly **manoeuvrable**, using their **dorsal** and **anal** fins to swim, and can even swim backward.
- Triggerfish get their name from their first three dorsal spines. They use the large first spine to lock themselves into rocky crevices at night or when threatened. This spine can only be unlocked by depressing the smaller second spine - the "trigger".
- Their **gills** are not visible slits like other fish; they are tiny openings hidden behind their tough, armour-like scales.
- Triggerfish eat sea urchins, crabs, and worms. To eat a sea urchin, they will flip it over to attack its less-spiny underside. They also blow jets of water from their mouths to uncover buried prey in the sand.
- When protecting their eggs, triggerfish are incredibly **aggressive** and known to charge, bite, or chase divers and other intruders. Females will also blow water on their eggs to keep them oxygenated.
- Unlike most fish, they swim primarily by **undulating** their **dorsal** and **anal** fins, allowing them to hover, move forward, or move backward with great precision.
- The Picasso triggerfish, or Humuhumunukunukuapua'a, is the state fish of Hawaii and is famous for its **vibrant**, Picasso-like painting patterns.
- They are intelligent fish that learn from their surroundings and have been observed interacting with their **environment** in calculated ways.
- They have incredibly strong jaws and teeth designed to crush shells and, in some cases, crack open sea urchins.
- While generally **solitary**, several species of triggerfish time their **spawning** activities to the moon and tides, often gathering in large numbers during new or full moons.
- Some species, like the Giant Triggerfish, can grow to over 0.9m long.

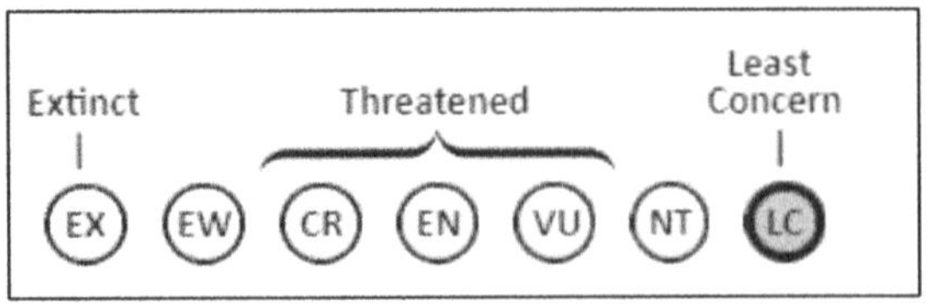

[1]

Illustrated are:
Wedgetail triggerfish (*Rhinecanthus rectangulus*) also known as reef triggerfish, rectangular triggerfish
Lagoon triggerfish (*Rhinecanthus aculeatus*) also known as blackbar triggerfish, Picasso triggerfish

Sooty albatross

Phoebetria fusca

- Sooty albatrosses are seabirds with a striking appearance and interesting behaviours.
- The sooty albatross is also known to sailors as 'the Quaker'.
- They have a dark, sooty-brown to black plumage, a white ring around their eye, and a black beak with a yellow or orange stripe. T
- These birds are known for their **impressive wingspan**, which allows them to glide effortlessly over the ocean for long distances.
- Sooty albatrosses are medium-sized albatrosses, with a wingspan of about 2m.
- Their **plumage** is a dark, sooty-brown to black colour. The head and sometimes the tail can be a darker shade than the rest of the body. They have a **distinctive** white ring around their eye and a yellow or orange stripe on their beak.
- In general, the sooty albatross can be found **foraging** over the deep ocean. Non-breeding adults do not often venture out of their native oceans but have a broader **foraging** range, while breeding adults **forage** closer to their breeding sites (within 600km) and **juvenile** birds tend to be confined to **subtropical** waters.
- They eat fish, **crustaceans**, squid, and **carrion**. They often **forage** alone, but may join mixed-species **flocks** at times, especially when following fishing boats.
- Sooty albatrosses are masters of the wind. They use their long **wingspan** to glide for long distances, often without flapping their wings, making them very efficient flyers.
- They breed in loose **colonies** on cliffs or steep slopes. A single egg is laid in October or November, and both parents help to **incubate** it and raise the chick.
- Like all albatrosses, sooty albatrosses drink saltwater. They have special glands above their eyes that help them get rid of the excess salt, which can sometimes drip from their nostrils.
- Sooty albatrosses can live for a long time, with some individuals living for over 60 years.
- Sooty albatrosses are currently listed as near threatened due to threats like **bycatch** in fishing gear and climate change.

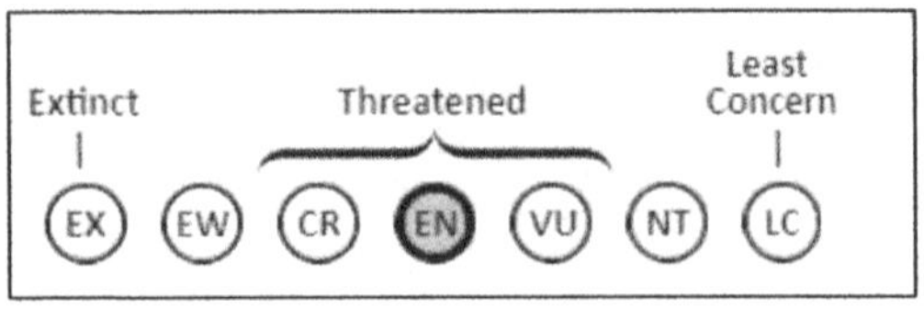

[1]

EPBC List = V =Vulnerable

Yellow-bellied glider

Petaurus australis

- Yellow-bellied gliders are fascinating **marsupials**. They have a black stripe down their back and tail, and their belly fur can range from white to yellow, with older gliders having more yellow fur.
- They are **social** creatures, living in family groups and defending large **territories**. Their diet includes **nectar**, sap, and insects.
- They are gliding experts having a special **membrane** of skin called a **patagium** between their wrists and ankles that stretches out, allowing them to glide through the air like a flying squirrel. They can glide up to 100m in one jump, and some have been recorded gliding even further.
- They are highly **adapted** to life in the trees, spending almost all of their time in the forest **canopy**.
- They create V-shaped incisions in the bark of eucalypt trees to access sap, a key part of their diet. Yellow-bellied gliders are highly selective about which trees they feed on and may revisit the same trees for over 20 years. Their diet includes tree sap, **nectar, honeydew, pollen**, and insects.
- They live in family groups, with males, females, and their young, and these groups defend large **territorie**s.
- Yellow-bellied gliders are nocturnal and vocalise with loud shrieks and calls, especially at night, that can be heard up to 500m away. Their calls serve both communication within family groups and **territorial** defence.
- They rely heavily on eucalypt trees for sap and also utilise hollows in these trees for denning and breeding. They require hollows in trees for shelter and breeding, making the preservation of old-growth forests crucial.
- **Habitat loss** and **fragmentation** due to deforestation and timber harvesting are major threats to their survival.

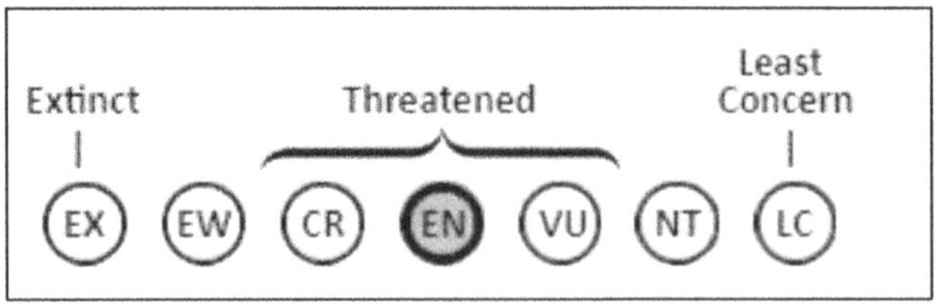

[1]

Black-tailed godwit

Limosa limosa

- There are four recognised **subspecies** of the black-tailed godwit, all with orange head, neck and chest in breeding **plumage** and dull grey-brown winter colouration, and a distinctive black and white wingbar at all times.
- Known for their long **migrations**, these elegant wading birds breed in wet grasslands and bogs in places like Iceland and Siberia, but they spend their winters in warmer areas like southern Europe, Africa, and Asia, **migrating** in **flocks**. They are occasional but uncommon visitors to the GBRMP.
- Black-tailed godwits are tall birds with long legs that are great for wading in shallow water and a long, slender beak 7.5-12cm long, that they use to **probe** the mud for tasty insects, worms, and small creatures.
- In flight, its bold black and white wings and white **rump** can be seen readily. When on the ground, it can be difficult to separate from the similar bar-tailed godwit, but the black-tailed godwit's longer, straighter **bill** and longer legs are **diagnostic**. Black-tailed godwits are similar in body size and shape to bar-tailed godwits but stand taller.
- The most common call is a **strident** *'weeka weeka weeka'*.
- They are amazing travellers! They fly long distances from their breeding grounds to their wintering areas, often in large groups.
- These birds are very loyal. They often form strong bonds with their mates that can last for many years, returning to the same breeding grounds each spring to raise their chicks together. Both the male and female black-tailed godwit work together to **incubate** the eggs and take care of the chicks after they **hatch**.
- A study of the Icelandic population showed that despite spending winter apart, pairs are reunited on their breeding grounds within an average of three days of each other. If one partner does not arrive on time, 'divorce' occurs. They nest in loose **colonies**. Unpaired males defend a temporary **territory** and perform display flights to attract a mate. Several nest scrapes are made away from the **courtship territory** and are defended from other godwits.[20]
- The species breeds in fens, lake edges, damp meadows, moorlands and bogs and uses estuaries, swamps and floods in the Northern Hemisphere winter; it is more likely to be found inland and on freshwater than the similar bar-tailed godwit.

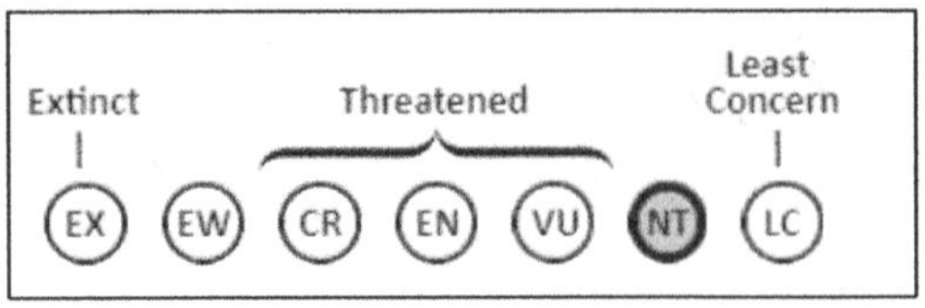

[1]

EPBC List = E =Endangered

Red knot

Calidris canutus

- Red knots are shorebirds known for their incredible long-distance **migrations**, travelling thousands of miles between their Arctic breeding grounds and wintering sites in places like Australia and South America.
- There are six **subspecies** of red knots with the Eurasian populations wintering in Africa, Papua New Guinea, Australia and New Zealand.
  Red knots are among the longest-distance **migrants** in the animal kingdom, flying over 14,000km each way between their breeding and non-breeding areas.
- Despite being small birds with a wingspan of only about 47-53cm, they can fly non-stop for very long distances.
- Their diet varies according to season. On the breeding grounds, knots eat mostly spiders, **arthropods**, **larvae**, and some plant material obtained by surface pecking. Red knots love to eat **bivalves** like clams and mussels, which they find by **probing** the mud and sand at low tide. They eat the shells whole, and their strong **gizzards** crush them to digest. They also eat worms and **crustaceans**.
- This species forms enormous **flocks** when not breeding. They often gather in large, tight groups, especially on sandy beaches and mudflats, and can even intermingle with other shorebirds like great knots.
- Red knots change colour! Their feathers turn a rusty red when they head north to breed, then return to a more greyish colour for their wintering journey.
- They breed in cold climes, including the Arctic tundra, Greenland, the Taymyr Peninsula, the Chukchi Peninsula and some Siberian islands, building shallow nests on the open ground with their eggs **camouflaged** with blotches and spots.
- During their long **migrations**, and in the case of the North American population, places like the Delaware Bay are vital rest stops for them, where they feed on horseshoe crab eggs to get ready for the final leg of their journey.
- Red knots are facing challenges due to climate change, which could affect their Arctic breeding grounds and the availability of important food sources.

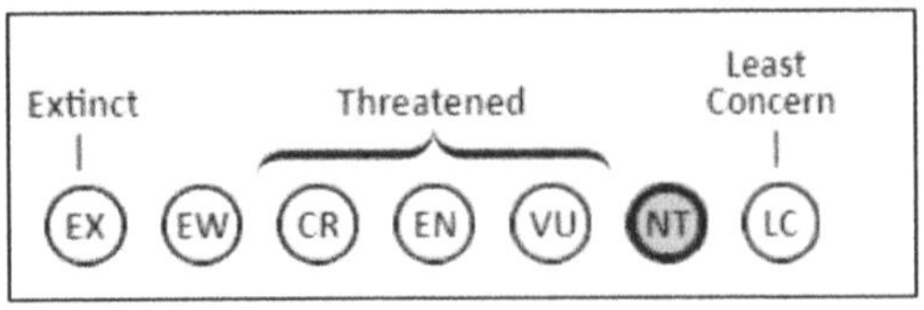

[1]

EPBC List = E =Endangered

Brolga

Grus rubicunda

- Brolgas are large, graceful birds found in Australia and New Guinea, known for their impressive size, unique dances, and their special place on the Queensland coat of arms.
- In northern Queensland, they can easily be confused with the Sarus cranes, which have more extensive red on their heads and necks, and pinkish legs. To further confuse identification, two **hybrid species** have emerged in northern Australia called the Sarolga or Antigone.
- Brolgas are one of Australia's largest flying birds, standing up to 1.4m tall with a wingspan of over 2m. They have a pale grey body, long legs, and a distinctive red-to-orange patch on their head, with a black "dewlap" (hanging skin) under their chin.
- Brolgas are famous for their elaborate **courtship** dances, which involve leaping, head shaking, and loud trumpeting calls, helping to strengthen their pair bonds. These dances are so **iconic** that they are featured in Aboriginal legends and dances.
- You can find brolgas across tropical northern Australia. There is a significant population in Victoria and New South Wales, though their conservation status varies by region.
- They inhabit shallow open wetlands, coastal mudflats, and irrigated farmlands. They need these wetland areas for feeding and breeding.
- Brolgas are **omnivores**, meaning they eat both plants and animals. Their diet includes wetland plants like sedge **tubers**, as well as insects, **amphibians**, and even small animals like mice. They are very skilled at **foraging**, even with their heads submerged underwater.
- The brolga is a significant symbol for Queensland, as it was added as a supporting animal to the state's coat of arms in 1977 and officially named the state bird in 1986.
- Brolgas are thought to mate for life, and both parents work together to **incubate** eggs and care for their chicks for up to a year or more.
- Brolgas are **unique** among cranes as they possess a gland near their eyes that helps them get rid of excess salt, allowing them to live near saltwater marshes and even drink **saline** water. This specialised salt gland is not found in Sarus cranes, which appear to rely on freshwater sources.
- Brolga populations have declined due to **habitat loss** and **modification**, with specific conservation efforts underway.

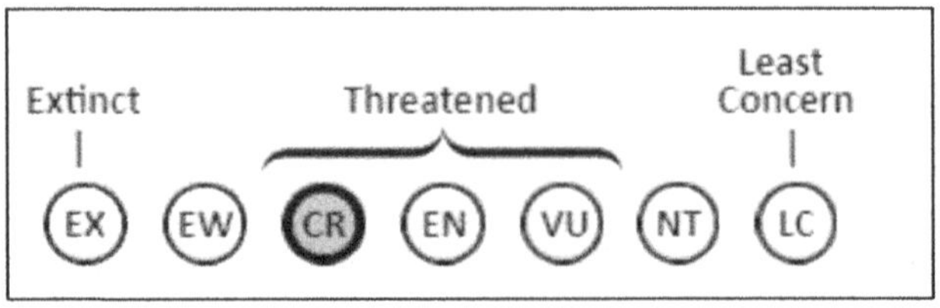

[1]

Ruddy turnstone

Arenaria interpres

- These birds are found along coastlines worldwide and migrate to breeding grounds in the Northern Hemisphere, then travel south for the non-breeding season, including to Australia from September to May.
  Ruddy Turnstones get their name because they use their bills to flip over stones and seaweed to find yummy insects, **crustaceans**, and other small creatures hiding underneath.
- They are global travellers and migrate very far! Some Ruddy Turnstones fly from Australia to Taiwan in just about six days, covering thousands of miles. They are capable of impressive non-stop journeys of up to approximately 7600km.
- They are amazing divers and strong fliers and can even dig holes in the sand to catch **prey** that's buried deeper.
- Their diet is varied, including insects, spiders, **crustaceans, molluscs**, and sometimes even eggs or **carrion**. They will eat maggots in rotting seaweed and sandhopper **larvae**.
- Ruddy Turnstones have different looks for different times of the year. They are chunky birds with short orange legs and bills, and distinctive plumage that changes between breeding and non-breeding seasons. In breeding plumage, males have a bright reddish back, black chest patches, and a black and white face, while females and winter plumage are duller.
- Their calls include a staccato "tuk-a-tuk" or a warbling "tuk-tuk-tuk-tuk".
- Ruddy Turnstones are **monogamous** and usually return to the same mates and breeding grounds each year. Both parents care for the chicks, though the female departs before **fledging**.
- Ruddy Turnstone chicks are tough - they can leave the nest soon after **hatching** and are able to feed themselves, though they can fly after about a month.
- Outside of breeding season, they often gather in large groups, sometimes mixing with other types of shorebirds, but they can be quite **aggressive** when competing for food or defending their **territory** during breeding season.

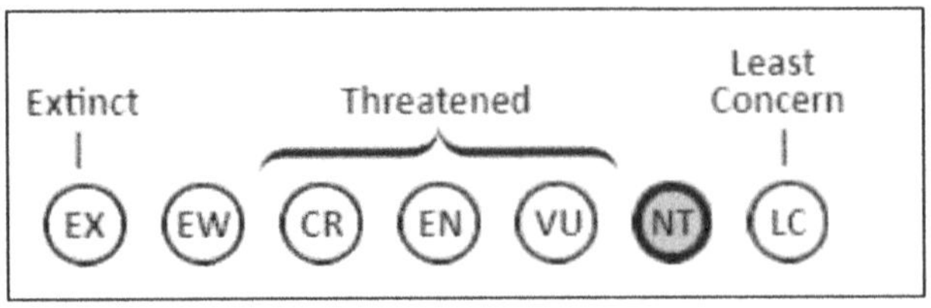

[1]

EPBC = V = Vulnerable

Western Alaskan bar-tailed godwit

Limosa lapponica baueri

- Bar-tailed godwits are large wading shorebirds with a length of 37-39cm, a wingspan of 62.75cm, and a weight between 250-450g. They have a long neck and a long, upturned **bill** with a dark tip.
- Bar-tailed godwits do not breed in Australia but come here from Scandinavia, northern Asia, and Alaska for the winter. Satellite tracking has confirmed that some birds fly over 11,000km, and one tracked godwit flew 29280km in a round trip lasting 174 days!
- The **migration** of the western Alaskan bar-tailed godwit across the Pacific ocean is the longest known non-stop flight of any bird and also the longest journey without pausing to feed by any animal.
- Scientists are still trying to figure out exactly how they sleep during their long flights, but they think they might take very short "microsleeps" while gliding.
- To fly for so long without eating, the Bar-tailed Godwit can shrink its digestive organs, like its stomach and intestines, to save weight and make room for extra fat for the long flight. Their breeding **plumage** is a distinctive rusty red colour.
- They feed on bristleworms and **shellfish** on coastal mudflats, beaches, mangroves, and estuaries, probing their long bills rapidly into the shallows or exposed mud.
- Feeding parties can be 30 or more birds, including **migratory** adults and some birds that remain all year round, mostly young birds.
- If you see a group of Bar-tailed Godwits together, you can call them a "prayer" or an "omniscience".
- Threats include **habitat** loss and **degradation**, **overexploitation** of **shellfish**, **pollution/contamination** impacts, diseases, extreme weather events, and climate change, which can raise sea levels.

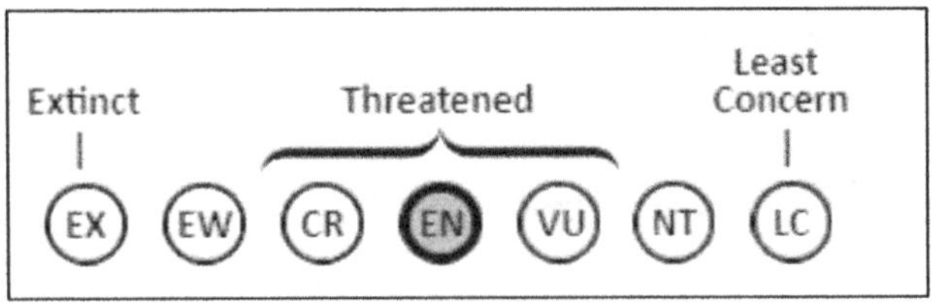

[1]

EPBC list = E = Endangered

Grey-headed flying fox

Pteropus poliocephalus

- The grey-headed flying fox is amongst the largest bats in the world. Adults can weigh up to 1kg (body weight ranges from 120–1600g). Males are usually bigger than females.
- and have a distinct reddish-brown neck collar with a light grey head.
- They are commonly known as fruit bats or flying foxes and are also the largest bats in Australia.
- They eat fruits and other plant matter, occasionally also eating insects, which they can find due to their keen sense of smell.
- They are mostly **nocturnal**.
- They are the only flying-fox species with fur extending all the way down their legs to their toes.
- Six flying fox species have already been made **extinct** (due to overhunting)!
- They are important **pollinators**.
- Unlike many bats, they do not use **echolocation**. They have large eyes and excellent vision and smell, which they use to find food at night.
- They use their large wings as fans to cool themselves on hot days, with **colonies** sometimes appearing to "flutter" in unison.
- They travel up to 750km between seasonal camps to follow the flowering of native trees like eucalypts.
- Mother bats leave their pups in "**creches**" at their camps while they **forage**, returning at dawn. They are long-lived **mammals** with an average reproductive age of 6–10 years and typically produce only one pup per year.
- They can also carry diseases such as Australian bat Lyssavirus, rabies, Hendra virus, and Nipah virus.
- They have **cultural significance** to **indigenous** people, appearing in traditional art, folklore, and weaponry. Their fur and teeth were used as **currency** in the past, and some **cultures** still use them as **currency**.
- In terms of threats, flying foxes' **predators** include sea eagles and snakes.
- They also suffer from habitat loss for **foraging**, competition, and extreme weather events. Recent studies indicate that since 1994, over 24,500 grey-headed flying foxes have died due to extreme heat events.

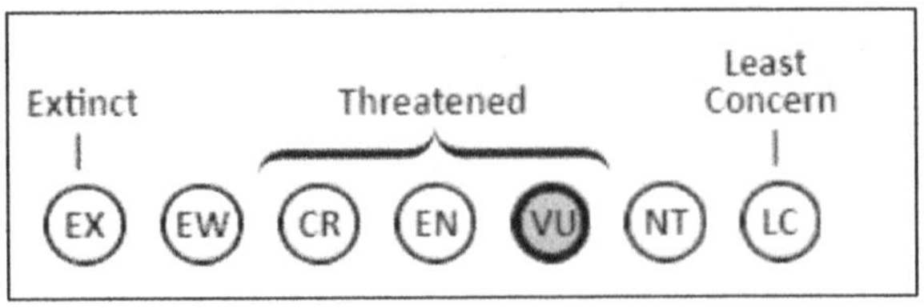

[1]

EPBC List = V = Vulnerable

Grey falcon

Falco hypoleucos

- The grey falcon is a **rare** and **elusive** medium-sized **raptor** found in **arid and semi-arid** Australia, known for its fast aerial hunts and distinctive pale grey and white **plumage**.
- Grey falcons are a compact falcon, mostly grey on top with white undersides and have a yellow **cere** (the fleshy part above the beak) and legs. When they fly, the tips of their wings are darker.
- Adult grey falcons are about 30-45cm long with a wingspan of 85-95cm. It is notable for its reversed **sexual dimorphism**, with females being significantly larger than males (30%), and for being a specialist **predator** of birds, often hunting in aerial pursuits.
- They prefer to live in **arid and semi-arid** environments, open areas with sparse trees in inland Australia, usually where there is less than 500mm of rain each year and near watercourses.
- While they don't **migrate** in the traditional sense, their movements are **nomadic**, driven by the need to find resources and avoid deteriorating conditions.
- They are known for their **agile** and fast flight.
- They are amazing at catching prey by chasing them through the air very quickly. The grey falcon primarily **preys** on other birds, especially parrots and pigeons, and hunts them during fast, level aerial pursuits close to the ground. They also consume small **mammals, reptiles**, **and invertebrates** like locusts.
- You can often hear their calls, which are a clucking or chattering sound.
- The grey falcon is one of the world's **rarest** falcons, and the **rarest** of the six falcons found in Australia, so seeing one is very special.
- Grey falcons don't build their own nests! They use nests made by other birds, like large trees near water.
- You might sometimes see a grey falcon family, which includes an adult pair and a young bird.
- Like many animals, they face threats from things like **feral** cats and changes in their environment, like increased temperatures due to climate change.

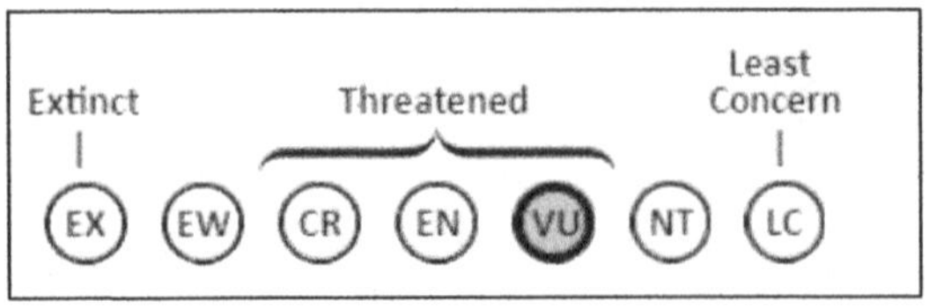

[1]

EPBC List = V = Vulnerable

Lesser sand plover

Charadrius mongolus

- The lesser sand plover, also called the Mongolian plover, is a small bird found on coastal areas like beaches and mudflats in Australia, known for its **impressive** long-distance **migrations** and a **unique** feeding method to catch insects, **crustaceans**, and worms. They are highly **social** birds, often seen in large **flocks**, especially when **roosting** during high tide, and sometimes mingle with other shorebird **specie**s.
- Lesser sand plovers are amazing long-distance travellers, **migrating** to Australia for the summer and then flying back to other parts of the world for their breeding season, primarily to Siberia.
- They don't just peck for food; they have a special "run-stop-peck" dance where they run, stop, and then quickly lunge to catch yummy insects, worms, and little crabs hiding in the sand or mud.
- These birds love to hang out with other plovers and shorebirds, forming big groups of over 100 individuals, particularly when **roosting** on sandy beaches, spits and rocky shores.
- You'll most likely see them at the beach, especially on the **intertidal** sandflats and mudflats, which are areas of the beach that get covered by water at high tide and are exposed at low tide.
- Like other plovers, they might use a "false brooding" display to trick **predators** away from their nests if they get too close. They might pretend to be hurt or have a broken wing to lure the danger away from their precious eggs or chicks.
- When lesser sand plover babies **hatch**, they are born with their feathers and can already walk and follow their parents! They can even join their first **migration** within a few weeks.
- The flight call of the lesser sand plover is described as a hard trill.
- Like other shorebirds, they are sensitive to changes in their **habitat**, so it's important to keep beaches clean and protect their nesting areas.

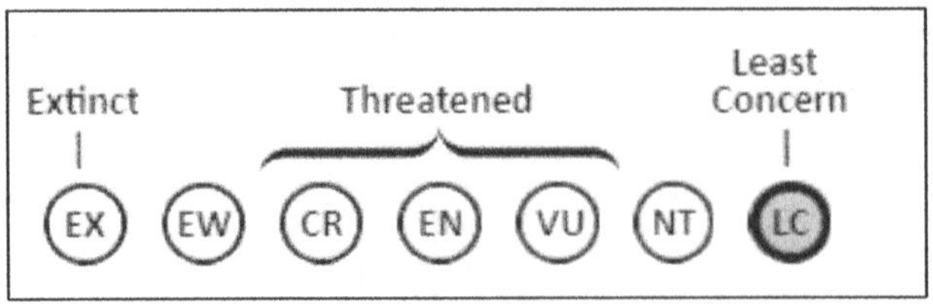

[1]

EPBC List = E = Endangered

Wrasse species

Family: Labridae

Illustrated here are six species, all of which have been assessed on the IUCN Redlist:

- **Checkerboard wrasse** *(Halichoeres hortulanus)* – IUCN Redlist = LC
- **Clown wrasse** *(Coris gaimard)* – IUCN Redlist = LC
- **Diamond wrasse** *(Anampses caeruleopunctatus)* – IUCN Redlist = LC
- **Leopard wrasse** *(Macropharyngodon meleagris)* – IUCN Redlist = LC
- **Moon wrasse** *(Thalassoma lunare)* – IUCN Redlist = LC
- **Three ribbon wrasse** *(Stethojulis trilineata)*– IUCN Redlist = LC

- Wrasses are a **diverse** family of marine fish known for **vibrant** colours, **protogynous hermaphroditism** (starting life as female and transforming to male), and **protrusible** teeth. There are over 600 species in 81 genera, which are divided into eight subfamilies.
- They are typically small, most of them less than 20cm long, although the largest, the humphead wrasse, can measure up to 2.5m.
- They range from small reef cleaners to the massive humphead wrasse, often acting as **crucial** pest controllers on reefs.
- Most wrasses begin life as females and can turn into males later, often triggered by the loss of a **dominant** male.
- The cleaner wrasse (*Labroides dimidiatus*) has passed the mirror test, suggesting it is self-aware. They are known for establishing "cleaning stations" where they remove **parasites** and dead tissue from larger fish, even inside their mouths.
- Wrasses have **protrusible** teeth, which means they can project their jaws outward to catch prey quickly.
- Some of the smaller wrasses follow the feeding trails of larger fish, picking up **invertebrates** disturbed in their passing.
- **Juveniles** of some genera hide among the tentacles of the free-living mushroom corals and *Heliofungia actiniformis*.
- Many wrasse species burrow into the sand to sleep at night to protect themselves from **predators**.
- They are often quite **curious** and interactive with divers, with some individuals even allowing themselves to be nudged or rubbed.

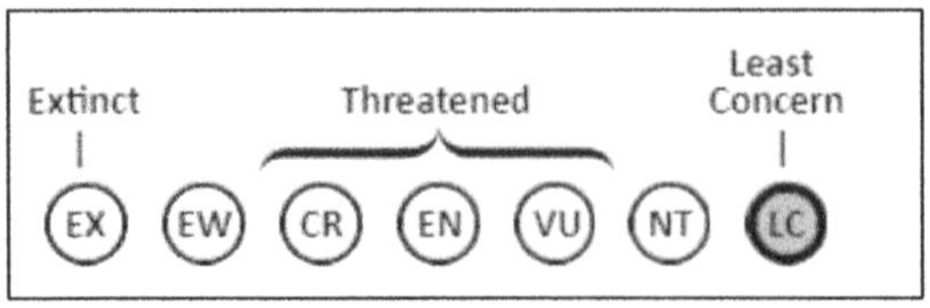

[1]

Bluestriped snapper, blue-lined snapper

Lutjanus kasmira

- The bluestripe snapper or blue-lined snapper is a **vibrant** reef fish widely distributed across the Indo-Pacific. Known for its brilliant yellow body and thin blue stripes, this species has several fascinating **biological** and behavioural **characteristics**.
- They possess a bright yellow body with four narrow, brilliant blue **horizontal** stripes running along their sides. The belly is typically white, and the fins are a bright yellow.
- While highly colourful in the wild, the blue stripes can fade, and the yellow may become less intense when they are kept in aquariums or as they age.
- **Juveniles** are often **distinguished** by a blackish band or spot that runs from the tip of the snout to the eye, sometimes with a dark spot under the dorsal fin.
- These fish are highly **gregarious**, living in large **schools** (shoals) around coral reefs, caves, and wrecks during the day.
- They are often found resting in stationary, large groups around coral "bommies" (coral heads) during the day.
- They are **carnivorous** and feed on fish, **crustaceans** (shrimps, crabs), **cephalopods**, and **plankton**.
- Their eggs are **planktonic** and **hatch** very quickly—within just 18 hours. They **spawn** in pairs that briefly separate from the main **school**. The pair swims in a spiral towards the surface to release eggs. While they can reach up to 40cm in length, they are more commonly found around 20–25cm.
- The species typically lives for about 6 years, with some studies suggesting they reach **maturity** around 3 years of age.
- They are found throughout the Indian Ocean and the Pacific Ocean, from the coast of Africa to the Marquesas Islands, including the Red Sea and Australia.
- They prefer coral reef **environments**, **inhabiting** both shallow **lagoons** and outer reef slopes at depths ranging from 1 to 265m, though they are usually found above 60m.
- When introduced outside their **native** range (e.g., in Hawaii), they can become **invasive**, outcompeting native fish for food and **habitat**.
- They are considered a popular food fish and a game fish in many parts of their **native** range.
- This snapper species, with its striking yellow and blue colouring, is a **vibrant** part of the **tropical** reef **ecosystem**.

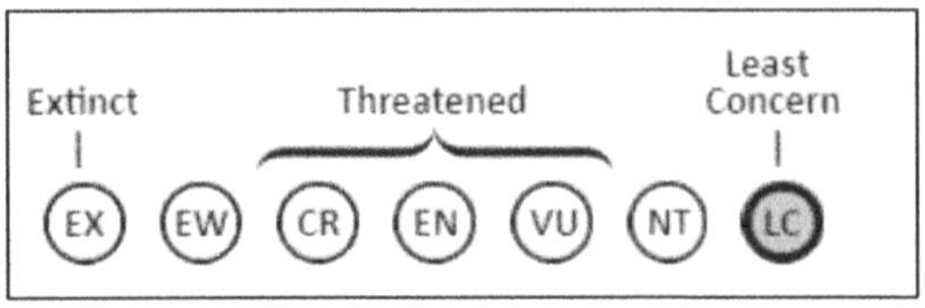

[1]

Harlequin tuskfish

Choerodon fasciatus

- The Harlequin tuskfish is a visually striking **marine** wrasse **native** to the Western Pacific. It is famous for its tiger-striped pattern, **protruding** bright blue **canine** teeth, and **unique predatory** habits.
- Its scientific name, *Choerodon*, comes from the Greek words for "pig" and "tooth". It features four highly noticeable **canine** teeth that turn a **vivid,** electric blue as the fish matures.
- Adults have 7 to 8 pairs of **vertical,** alternating orange/red and white bands, edged in **iridescent** blue.
- **Specimens** found near Queensland, Australia, tend to feature much brighter, richer red and blue hues.
- Tuskfish are remarkably intelligent and use rocks as "anvils". They are known to grab hard-shelled **prey** and repeatedly smash them against hard surfaces to break them open.
- Their large eyes can move independently of each other, allowing them to survey their entire environment for **predators** or food at a glance.
- Younger tuskfish are paler and have large black spots, known as **ocelli**, on their fins. These false eyespots confuse **predators** and disappear as the fish ages.
- Like many members of the wrasse family, all Harlequin tuskfish hatch as females. If a dominant male is lost, the largest, most dominant female will naturally transition into a male to replace it.
- They inhabit **coastal** coral reefs, **lagoons**, and reef slopes at depths ranging from 5 to 35 metres. They typically spend their days hiding in reef crevices and hunting for **benthic invertebrates** like crabs, sea urchins, and snails.

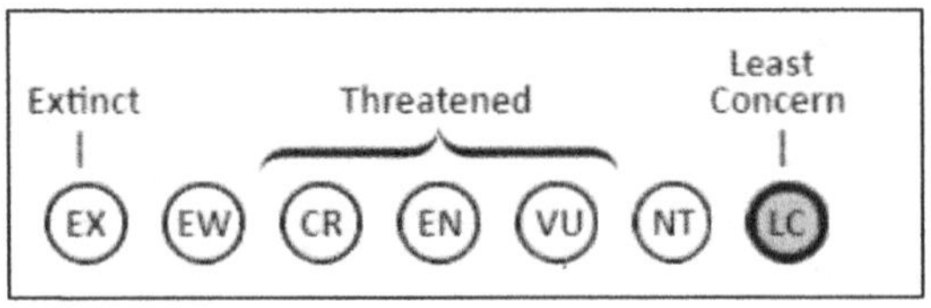

[1]

Tropical two-wing flyingfish

Choerodon fasciatus

- The tropical two-wing flyingfish is a master of aerial evasion. Growing to a common length of 20cm, it launches itself out of the water at high speeds, spreading exceptionally large **pectoral** fins to glide for long distances and escape swift ocean **predators**.
- They don't flap their fins like birds; instead, they use their powerful tails to accelerate up to 60km/h underwater before launching into the air, where their stiff, wing-like **pectoral** fins keep them airborne.
- They can glide for distances well over 200m, typically staying aloft for a few seconds to **evade** hunters like tuna and dolphinfish.
- Unlike other flying fish species that have enlarged **pelvic** fins (acting as "four wings"), *Exocoetus volitans* uses only its two enlarged **pectoral** fins.
- To blend in from both above and below, they are **iridescent** dark blue on their backs and silvery-white on their bellies.
- Despite their small size, they have excellent eyesight and hunt **individual planktonic crustaceans** rather than mindlessly filtering water.
- Rather than attaching their eggs to floating seaweed or **debris**, females release buoyant, **pelagic** eggs directly into open water.
- They are a **crucial** food source for large **pelagic** fish (such as swordfish) and seabirds.
- They have a maximum **lifespan** of approximately one year.

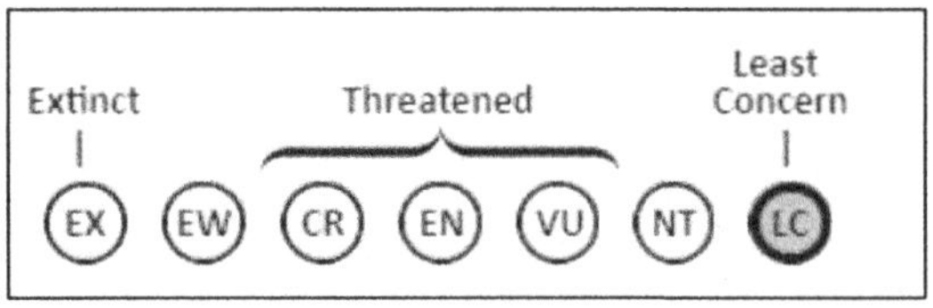

[1]

Humpback whale

Choerodon fasciatus

- The humpback whale is a **marine** giant famed for its haunting songs, impressive **acrobatic** leaps, and **epic migrations**. Growing up to 18m long and weighing around 40 tons, these gentle giants hold some of the most fascinating **traits** in the animal kingdom.
- Humpbacks undertake some of the longest yearly **migrations** of any mammal, swimming up to 8,000km between cold polar waters for feeding and warm **tropical** seas for breeding.
- Their scientific name literally translates to "big-winged New Englander" (Megaptera = big-winged, novaeangliae = New England). Their **pectoral** fins can grow up to 5m long—making them the longest flippers of any **baleen whale**.
- Humpbacks have bumps or **tubercles** on the head and front edge of the flippers; the tail **fluke** has a jagged trailing edge. The **tubercles** on the head are 5–10cm thick at the base and **protrude** up to 6.5cm. They are mostly hollow in the centre, often containing at least one **fragile** hair that erupts 1–3cm from the skin and is 0.1mm thick. The **tubercles** develop early in **gestation** and may have a sensory function, as they are rich in nerves. Sensory nerve cells in the skin are adapted to withstand the high water pressure of diving.
- Male humpbacks are famous for their long, **complex** songs that can last anywhere from 4 to 33 minutes and echo for miles underwater. Interestingly, all the males in a single region sing the same tune, but the song **evolves** and changes year to year.
- Their diet consists mostly of **krill** and small fish.
- When feeding, humpbacks use an **ingenious co-operative tactic** called "bubble-netting." A group of whales will swim in a circle beneath a school of fish and blow bubbles, which forces the fish into a tight, **disoriented** ball before the whales lunge up to swallow them in one massive gulp.
- Just like human fingerprints, the black-and-white patterns on the underside of a humpback's tail **fluke** (tail fin) are completely unique to each individual. This helps researchers track individual whales over their lifetimes.
- Humpbacks are incredibly **athletic** and frequently launch their entire 40-ton bodies out of the water in a behaviour known as "breaching". They also love "pec slapping" and "spy-hopping" to scan the surface.
- Orcas are the main natural **predators** of humpback whales.

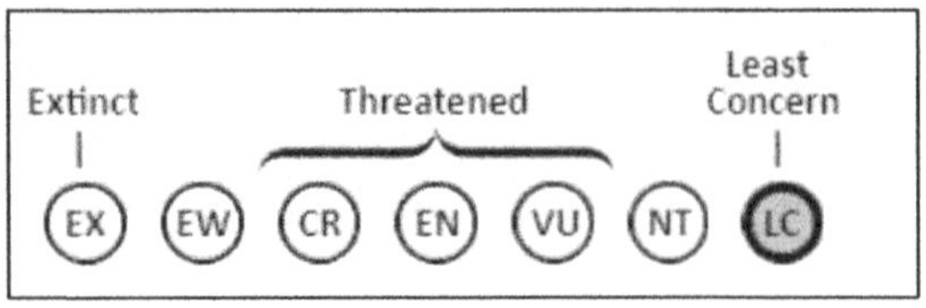

[1]

Oriental flying gurnard, purple flying gurnard

Dactyloptena orientalis

- The Oriental flying gurnard or purple flying gurnard is an unusual **marine** fish found across the **tropical** Indo-Pacific. Famous for its massive, wing-like **pectoral** fins and bottom-dwelling lifestyle, it is a fascinating and highly **adapted** creature of the seafloor.
- Despite its name, the flying gurnard does not actually fly or glide through the air like a flying fish. The name comes from the French word *"gurnard"*, meaning to grunt.
- Instead of swimming, it spends most of its time "walking" along sandy or muddy seabeds. It uses specialised rays in its **pelvic** fins and short **pectoral** fin rays to scuttle along the bottom in search of crabs, clams, and small fish.
- While hunting or if startled, it flares out its massive, rounded **pectoral** fins. These "wings" are not just for display; they feature **vibrant** blue, turquoise, or dark wavy spots designed to startle potential **predators**.
- Like its relatives, the sea robins, the Oriental flying gurnard possesses special "drumming muscles" attached to its swim bladder. It can vibrate these muscles to produce low, grunting or drumming sounds, particularly when captured or distressed.
- Unlike most streamlined fish, it has a robust, heavily armoured, box-like body covered in tough, protective bony scales. This makes it a tough target for smaller reef **predators**.
- Its base colouration is a grayish-brown covered with dark markings, allowing it to blend perfectly into sandy **estuaries** and shallow coastal bays.

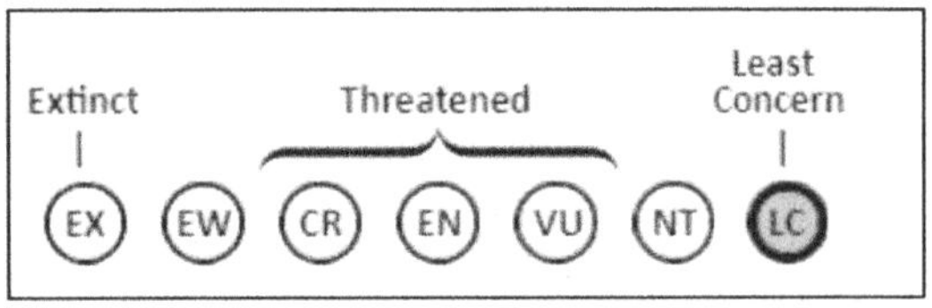

[1]

Shaggy anglerfish, shaggy frogfish, hispid frogfish, zebra anglerfish

Antennarius hispidus

- The shaggy or hairy frogfish is a master of **camouflage** and a highly **specialised ambush predator** found in the **tropical** waters of the Indo-West Pacific. It belongs to the anglerfish family and features a fascinating array of survival and hunting **adaptations**.
- Like other frogfishes, it has a modified **dorsal** spine that acts as a fishing pole. It wiggles an attachment at the end of this "rod" to lure in unsuspecting **prey**, often **mimicking** a small worm or swimming shrimp.
- Despite being almost entirely **sedentary**, its strike is one of the fastest in the animal kingdom. It can open its large mouth and suck its prey in as fast as 0.006 seconds, often swallowing fish nearly the same size as itself.
- Instead of swimming, it primarily "walks" along the ocean floor. It has **unique**, jointed **pectoral** and **pelvic** fins that act like little legs, allowing it to crawl around on rocky and muddy **substrates**.
- Its body is covered in small, hair-like **dermal spinules** that break up its **silhouette**. This allows it to blend perfectly into its surroundings, often **mimicking** clumps of **algae**, sponges, or coral rubble to hide from larger **predators**.
- It can shift its colouration to match its **specific** living **environment**. While it is usually light tan, yellow, or orange with narrow dark streaks, it can completely alter its **pigmentation** over several weeks.
- During **reproduction**, the female lays her eggs in a massive, **gelatinous mucous** raft that floats in the open ocean until the **larvae** hatch.

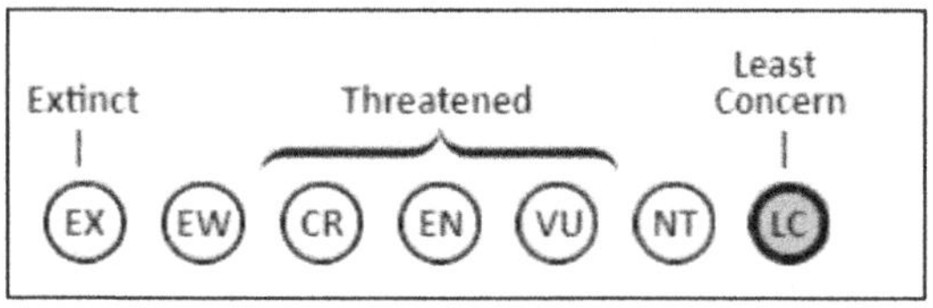

[1]

Spotted wobbegong

Orectolobus maculatus

- The spotted wobbegong is a master of **disguise** found in the temperate coastal waters of Australia, from Queensland to Western Australia. Easily recognised by their **unique** fringe-like "beards" and O-shaped spot patterns, these bottom-dwelling carpet sharks are **formidable nocturnal ambush predators**. They are **nocturnal**, resting during the day and feeding on fish and **invertebrates** at night.
- Their mottled patterns, flattened bodies, and fringing skin lobes (which give them their Aboriginal name, meaning "shaggy beard") allow them to blend perfectly with coral reefs and sandy ocean floors.
- They are **ambush predators** that lie completely motionless on the seafloor. To draw in **curious prey**, they can raise their heads and use the long **barbels** (whiskers) on their snouts as a living **lure**. When a fish swims too close, they snap it up and can even create intense suction to swallow large **prey** whole.
- It is a large, **robust** species, typically reaching 150–180cm in length.
- Unlike most sharks, they are remarkably flexible. If they need to, they can twist around and bite their own tails
- While they are capable swimmers, wobbegongs often use their strong, **lobed pectoral** and **pelvic** fins to waddle or walk slowly across the ocean floor.
- Many sharks must constantly swim with their mouths open to push water over their **gills**. Spotted wobbegongs have specialised openings behind their eyes called **spiracles**, which allow them to pull water over their **gills** while resting perfectly still.
- Like other sharks, their skin is covered in **sensory pores** that act as **electroreceptors**, allowing them to sense the electric fields of hidden crabs, octopuses, or fish.
- While not **aggressively** dangerous, they are known to have a powerful bite that can be hard to pry loose if they clamp down. Local divers have noted that they seem to remember being mistreated; there are anecdotal reports of them swimming up into the midwater to **harass** or bite a diver who had bothered them earlier.
- The spotted wobbegong was nominated for listing under the EPBC Act as a vulnerable species in 2013, but the nomination was rejected.

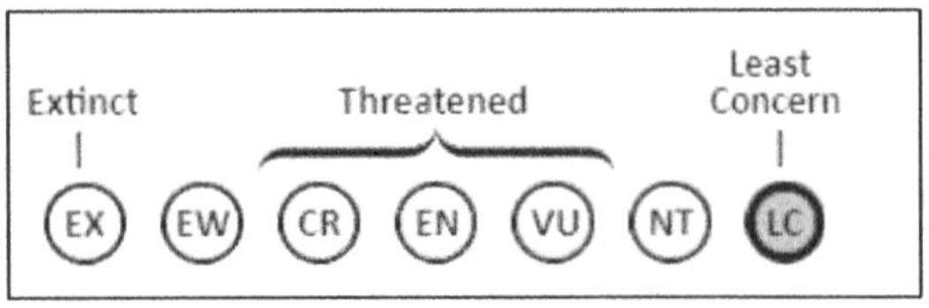

[1]

Peacock sole

Pardachirus pavoninus

- Sole fish are a family of highly **adapted** bottom-dwelling flatfish known for their **asymmetrical** bodies, masters of ocean-floor **camouflage**.
- Despite their delicate, flaky flesh, true soles are remarkably long-lived. The common sole (*Solea solea*) can live for up to 40 years. In many sole species (such as the English sole), the females significantly outgrow the males, often living up to 4 years longer.
- While some flatfish rely entirely on sight to hunt, many sole species are **nocturnal**. They rely heavily on their sense of smell and tiny **sensory filaments** lining their snouts to **forage** for worms, **molluscs**, and small **crustaceans** in complete darkness.
- The name "sole" comes from the Latin word *solea*, meaning "sandal" or "slipper". Because of their extremely flattened oval shape, they bear a striking resemblance to the sole of a shoe.
- The peacock sole is a fascinating flatfish found in the Indo-West Pacific, ranging from Sri Lanka to Japan and down to northern Australia. It is famous for its natural shark-repellent **mucous** and striking peacock-like eye spots.
- Both the Moses sole (*Pardachirus marmoratus*), found in the Red Sea, and the peacock sole release a special shark-repelling **mucous** when a shark attempts to bite them. These fish possess specialised glands along the bases of their **dorsal** and **anal** fins. When stressed, they secrete a milky, frothy fluid from these glands which is rich in **pardaxin**, a **potent peptide toxin**. **Pardaxin** acts as a powerful **gill** irritant to sharks. When a shark attempts to bite a Moses or Peacock sole, the **toxin** causes the shark's **gills** to flare uncomfortably, forcing the shark to instantly open its mouth and release the fish.
- As a newborn, the peacock sole looks like a normal fish with eyes on both sides of its head. As it **matures**, its left eye **migrates** to the right side of its head, and its body **compresses,** allowing it to lie flat on the seafloor while using both eyes to **ambush prey.** Adult soles swim and rest permanently on their blind, eyeless side.
- They are incredibly flat and oval, allowing them to easily bury themselves in sandy or muddy seabeds to hide from **predators**, leaving only their **protruding** eyes exposed.
- The skin on their "eyed side" contains **pigment**-changing cells that allow them to match the colour and texture of their immediate surroundings, functioning like an underwater cloak. Their 'blind' side is pale and flattened against the ocean floor.
- Its common name comes from the gorgeous pattern of reddish-brown spots and dark-ringed circles that resemble the eyes on a peacock's tail.

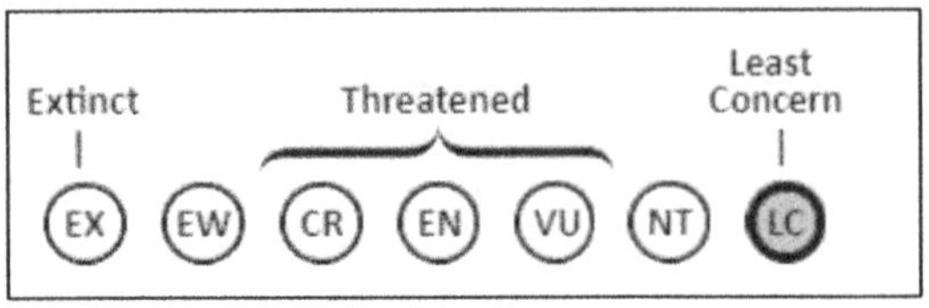

[1]

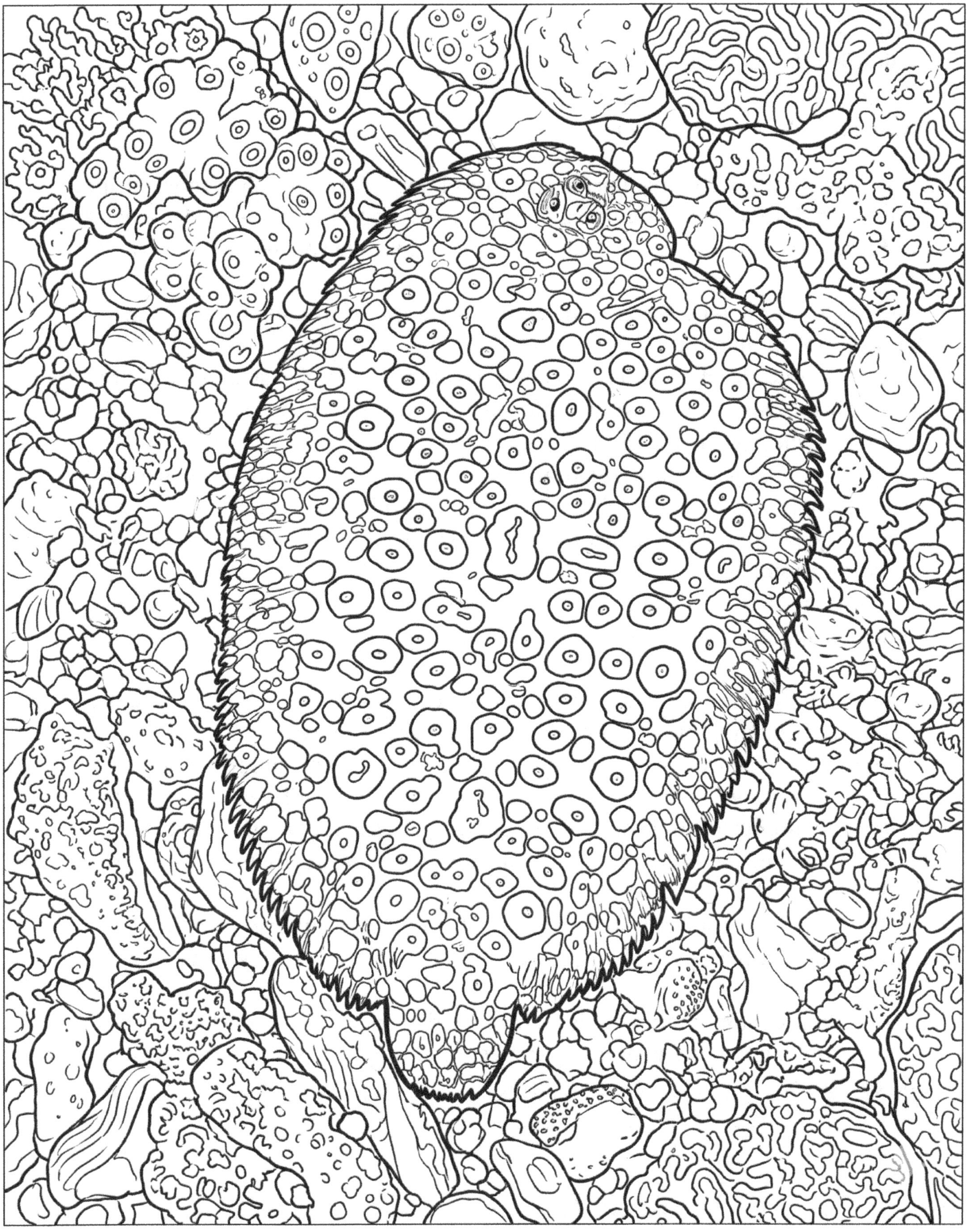

Lionfish species

Zebra lionfish *(Dendrochirus zebra)*
Spotfin lionfish *(Pterois antennata)*
Common lionfish *(Pterois volitans)*

- Lionfish are visually striking **marine** creatures known for their regal, zebra-striped bodies and **flamboyant**, **venomous** fin spines. Native to the Indo-Pacific, these **ambush predators** are infamous in the Atlantic as highly **invasive species** with **insatiable** appetites.
- They possess 13 needle-sharp, **venomous dorsal** spines that are used strictly for **defence,** not for hunting.
- They have **specialised cycloid** (smooth) scales and an adapted swim bladder that helps them stay **buoyant** while hovering effortlessly in the water.
- In the wild, they can live for 5 to 15 years.
- They use their large, fan-like **pectoral** fins to corner small fish and **crustaceans** in reef crevices, then swallow them whole in a single, lightning-fast gulp.
- While they naturally eat **crustaceans** and reef fish, when food is scarce, they will even resort to **cannibalism**.
- Because they spend a significant amount of energy swimming and hunting, they have **voracious** appetites, which makes them devastating to local reef **ecosystems** in non-native waters.
- Females are **prolific** breeders. A single female can release up to 30,000 **mucous**-filled eggs every few days, totalling around 2 million eggs per year.
- Accidentally introduced in the 1990s via the aquarium trade, lionfish have become a destructive **invasive species** in the Western Atlantic and Caribbean. They have very few natural **predators** in these regions, allowing their populations to explode.
- By eating **juvenile** fish and out-competing **native predators**, they severely threaten the **biodiversity** and health of coral reefs.
- While a sting from a lionfish spine causes extreme, localised pain (sometimes accompanied by nausea or fever), it is incredibly rare for the sting to be **fatal** to humans.
- Despite their **venom**, their meat is highly prized in some cuisines. When properly handled and filleted, the **venom** is harmless, and the flesh is described as flaky, tender, and delicious.

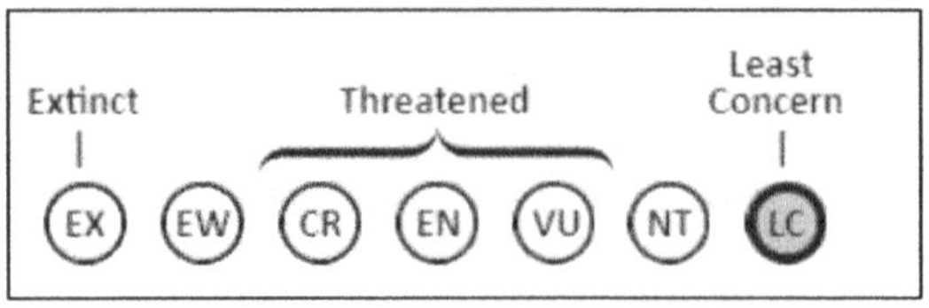

[1]

Common coral trout, Leopard coral grouper

Plectropomus leopardus

- The common coral trout is a **vibrantly** coloured **predatory** fish found in the western Pacific Ocean. Highly sought after in both **recreational** and **commercial** fisheries, they are **apex** reef **predators**, easily identified by their bright red, olive, or orange-red bodies, covered in regularly spaced small dark-edged blue spots, and a **distinct** blue ring around their eyes.
- Coral trout are incredibly effective hunters. They primarily eat reef fish (like damselfish) and use two distinct **tactics**:
    - **Ambush**: They lie perfectly still, blending into their coral surroundings, and lunge at passing prey.
    - **Prowling**: They slowly swim above the reef to **stalk schools** of fish higher up in the water column before striking with immense speed.
- They can also change their colour in seconds to create a patchy, mottled pattern while hunting.
- The common coral trout is a **protogynous hermaphrodite**. They all reach sexual **maturity** as females and then, as they grow older and larger, can transition into males. They generally start transitioning around 16 years of age and reach a size of about 32cm.
- During the new moon (most often between September and December), coral trout gather in large **spawning aggregations**. The exact timing is highly dependent on water temperature. After **courtship** behaviour—where males display their **dominance** and roll on their sides next to females—they release their eggs and **sperm** into the open water column to be **fertilised**.
- While they can live for up to 16 years, they experience rapid growth in their first three years of life, reaching the legal retainable size of around 38cm in about 2 to 3 years. They commonly measure around 35cm, but have been known to grow up to 120cm and weigh as much as 23.6kg.
- They are one of the most important components of the line fishery across the Great Barrier Reef and the wider Indo-Pacific. Because they fetch premium prices in the live reef fish trade, overfishing can sometimes deplete local populations around population centres.

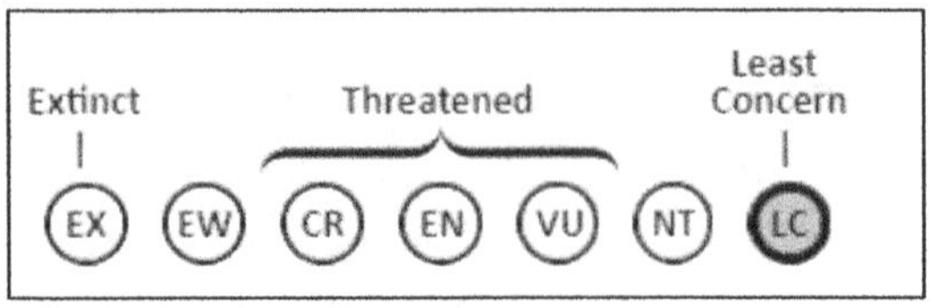

[1]

Japanese longtail dragonet

Callionymus japonicus

- The Japanese longtail dragonet, is a **marine** fish native to the western Pacific Ocean. A master of the seafloor, this dragonet is highly sought after by marine biology enthusiasts for its unique **physical traits** and behaviours.
- Males and females look completely different. Mature males sport remarkably elongated, flowing **dorsal** and **caudal** fins along with bright blue **longitudinal** stripes, while females are usually smaller, mostly brown, and **camouflaged** with darker blotches. While females grow to a maximum length of around 20cm, **mature** males can grow up to 30cm and develop remarkably elongated rays in their **dorsal** and **caudal** fins.
- These fish have flattened, triangular heads and have no scales. Their specific colour and mottling patterns provide perfect **camouflage** on coarse sand or gravel **substrates**, where they like to hide.
- While some dragonets live in shallow coral **lagoons**, this particular species generally prefers deeper, **subtropical demersal environments**, often found at depths of up to 200m.
- Despite their small size, they are highly efficient **micro-predators**. They spend their time on muddy or sandy bottoms sifting through the **sediment** for small **amphipods**, **polychaetes**, **bivalves**, snails, and **crustaceans**.
- Like other dragonets, they have fascinating, four-phase **courtship rituals**. **Territorial** males will **aggressively** defend their space and perform intense displays before pairing with females, then ascend into the water column together to release eggs and **milt**.

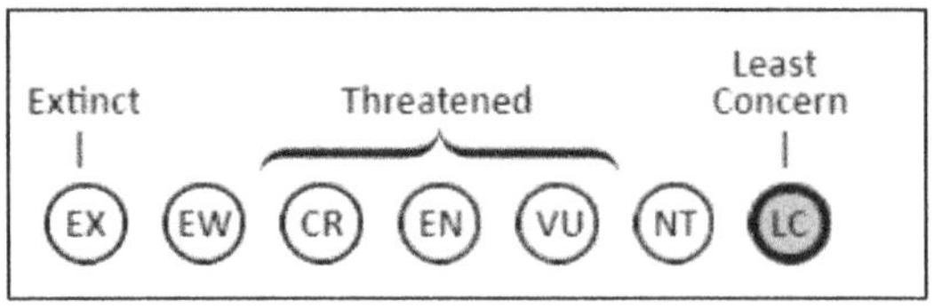

[1]

Spinner dolphin

Callionymus japonicus

- The spinner dolphin is a small oceanic dolphin famous for its **acrobatic** displays. Found in **tropical** waters globally, it gets its name from its ability to leap up to 3m into the air and spin **longitudinally** up to seven times in a single jump.
- They are champion spinners. Scientists believe this famous spinning behaviour helps remove **parasites**, acts as a form of communication, or simply helps them shed water and communicate across the surface.
- To avoid drowning, they cannot fully shut down their brains. They swim in long, slow, U-shaped patterns, resting one half of their brain at a time and periodically surfacing to breathe.
- Depending on the region, these dolphins live in **"fission-fusion" societies**. They gather in groups ranging from just a few dozen to several thousand to **socialise** and feed.
- During the day, they conserve energy in shallow, sandy-bottomed bays near oceanic islands where they can be observed **socialising** and **nurturing** calves.
- Their appearance can vary greatly depending on where they live. The most widespread **subspecies** is Grey's spinner dolphin, which features a **distinctive** three-banded colour pattern.
- Spinner dolphins exhibit highly **specialised**, **cooperative** hunting behaviours, but their **ecological** habits have also placed them directly in the crosshairs of **commercial** fishing.
- Spinner dolphins are highly efficient, **strategic** night hunters that **synchronise** their feeding with the movement of the ocean itself:
- During the day, their primary prey - small **mesopelagic** fish (like lanternfish), squid, and shrimp - hides in the dark ocean depths. At night, this massive **biomass migrates** upward toward the surface to feed on **plankton**. Spinner dolphins wait for this shift to hunt in shallower waters.
- When necessary, they are capable of diving to depths between 200 and 400m to seize **prey** with their sharp, pointed teeth.
- To maximise their catch, spinner dolphins hunt in organised groups. They use **echolocation** to track **prey** and swim in **coordinated** circles to pack fish into dense, tight patches. Pairs of dolphins will then take turns darting through the centre of the ball to feed.
- Spinner dolphins in the Great Barrier Reef Marine Park (GBRMP) face conservation threats from climate change, entanglement in fishing gear (**bycatch**), **marine debris**, and underwater noise **pollution**. While less vulnerable to **coastal** impacts than inshore species, they remain highly susceptible to offshore stressors

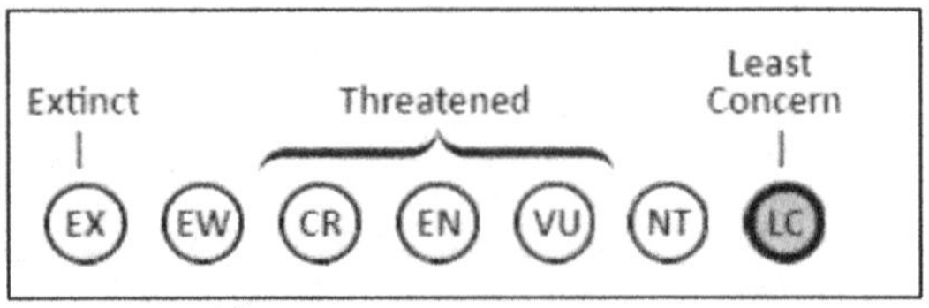

[1]

Indo-Pacific trumpetfish, Chinese trumpetfish

Aulostomus chinensis

- The Indo-Pacific trumpetfish is a master of **camouflage** and a highly skilled **ambush predator** found in **tropical** reefs. Known for its elongated, stiff body and long, tubular snout, this fascinating species uses **ingenious** "shadowing" techniques and explosive suction to hunt unsuspecting **prey**.
- They are famous for swimming directly alongside or behind larger, non-**predatory** fish (like parrotfish or groupers). By using these larger fish as moving blinds, they sneak up on smaller fish and **crustaceans** without triggering an escape response.
- Trumpetfish are relatively poor swimmers, so they rely entirely on **stealth**. They often hover perfectly still, facing head-down among **vertical** coral formations, sea whips, or sea fans, blending seamlessly into the **environment** until **prey** swims by.
- The tiny fleshy growth on the snout of the trumpetfish is a **barbel**. It contains **sensory** cells that help the trumpetfish detect nearby **prey** and navigate. It acts like a little fishing **lure** or an extra 'feeler' to catch small fish and shrimp off the reef.
- Once a target is within range, they utilise a "gape-and-suck" feeding **strategy**. When close to prey, their **tubular** mouths rapidly expand into a wide, circular gaping hole, creating a powerful **vacuum** that literally sucks in unsuspecting small reef fish and **crustaceans**.
- They can rapidly change colours to match their surroundings, shifting between uniformly bright yellow, mottled grey, dark green, or brown. They can also display pale **vertical** bars or **horizontal** stripes to break up their **silhouette**.
- Like seahorses and pipefish (to which they are related), their bodies are encased in stiff, interlocking rings of bone, making them quite **rigid**.
- If threatened by a larger predator, they can raise a series of short, sharp spines located along their back.
- In the Indo-Pacific, they are one of the few natural **predators** capable of feeding on **juvenile invasive** lionfish.
- The scientific name is rooted in Greek, combining *aulos* (meaning flute) and *stoma* (meaning mouth).

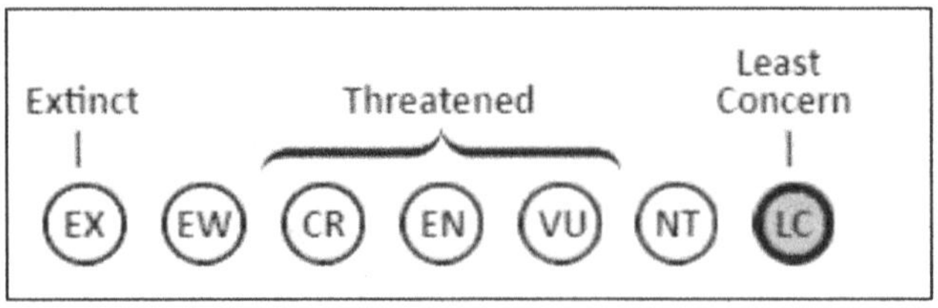

[1]

Beach stone curlew

Esacus magnirostris

- The beach stone-curlew is an **elusive**, masterfully **camouflaged** shorebird found along the coastline of Australasia and Southeast Asia. Weighing around 1kg and standing 55cm tall, it is one of the heaviest flying shorebirds in the world.
- Beach stone curlews do not migrate; they are strictly **sedentary**, **resident waders**. Once a breeding pair establishes a **territory**, they remain there year-round and can be incredibly site-faithful, often occupying the exact same stretch of coastline for over a **decade**.
- On a daily basis, their movements are governed entirely by the tides rather than the seasons. They **roost** in **coastal** mangroves or shaded **dune vegetation** during high tide and move down onto the exposed mudflats and sandbars at low tide to hunt.
- They have a distinctive, thick, and heavy **bill** that is black with a yellow base. They don't just **probe** the sand; they use this massive **bill** to **aggressively** spear and smash open hard-shelled **marine invertebrates** like crabs and **molluscs**.
- Unlike most birds that lay multiple eggs, they lay only one mottled, **cryptically** coloured egg directly onto the open sand or pebbles just above the high-tide mark. If the egg is lost to **predators** or a high tide, they will lay a replacement.
- While they are known for resting quietly during the day, they are largely **nocturnal** and **crepuscular** (most active at dawn and dusk). At night, breeding pairs will belt out a harsh, haunting, and high-pitched wailing sound.
- When threatened, instead of immediately taking flight, they will often squat and freeze to blend perfectly into the rocks and sand. If you get too close to their territory, they may **aggressively** march off while screeching.
- The chicks hatch in an advanced, highly developed state (**precocial**). However, they stay with their parents for an unusually long time, with some studies showing they remain dependent for 7 to 12 months before becoming completely independent.
- Because they spend their entire lives in saltwater environments with little to no access to fresh water, beach stone-curlews have specialised nasal glands that allow them to drink straight seawater and excrete the excess salt.
- They are rarely seen in groups. They live almost exclusively alone, in pairs, or as a small family unit (parents and their single chick). Their strictly **solitary** nature, combined with pale grey-brown **plumage**, makes them incredibly difficult to spot against standard beach gravel and sand.
- Because they depend entirely on undisturbed sandy beaches, sandbars, and **estuaries** for breeding and feeding, they are highly sensitive to human disturbance. They are classified as Near Threatened globally and are considered **critically endangered** in some regions (such as New South Wales) due to coastal development and beachgoers.

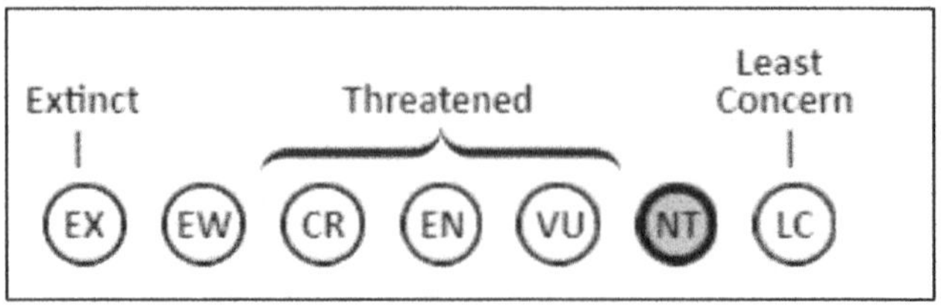

[1]

Foxface rabbitfish

Siganus vulpinus

- The foxface rabbitfish is a striking **marine** fish famous for its **vibrant** colouration, elongated "fox-like" snout, and incredible **algae**-eating abilities. Native to coral-rich **lagoons** and seaward reefs, it is a highly popular species in the **marine aquarium** hobby.
- Rabbitfish, named for their unique snouts and tooth plates, act as the ocean's gardeners, feeding primarily on **algae** to keep coral reefs healthy. However, they are highly capable **protectors**, equipped with painfully **venomous** spines in their fins.
- While many fish gather in massive, unstructured **schools**, many rabbitfish species are true romantics. They live and swim in **monogamous** pairs, staying together until death.
- In the case of the foxface rabbitfish, its common name comes from its narrow, elongated snout. This **unique** shape allows the fish to reach deep into reef crevices and staghorn coral branches to pluck out **algae** that other **grazers** cannot reach.
- About two-thirds of its body is a **vivid**, lemon-yellow, while the front portion—including the head and chest—features bold black and white banding
- Like all rabbitfishes (family *Siganidae*), the foxface possesses **venomous** spines on its **dorsal**, **anal**, and **pelvic** fins.
- While not **fatal** to humans, the **venom** can cause significant pain similar to a wasp or bee sting. It is primarily a **defence mechanism** used to deter **predators**, and humans are usually stung when accidentally handling the fish without protection.
- When startled, stressed, or sleeping at night, the foxface rabbitfish rapidly loses its **vibrant** yellow hue and turns dull, mottled grey or brown.
- When severely threatened by **predators** like eels, they have even been known to change to a dull colour and lie motionless, "play dead," then resume normal colour and swim away once the threat is gone.
- In the wild, they are roving **herbivores** that travel in pairs or **solitary** groups, constantly **grazing** on turf **algae** and **macroalgae**.
- By eating excessive **algae**, they prevent it from smothering and killing corals, making them essential to the reef's health and an incredibly effective cleanup crew for saltwater **aquariums**.
- Closely related to the *Siganus vulpinus* is the one-spot foxface (*Siganus unimaculatus*). It looks nearly identical, except for a distinct, single black splotch located just below the **dorsal** fin on either side of its body.

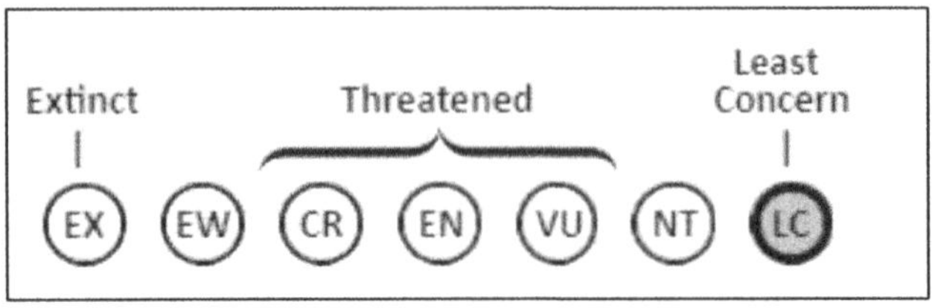

[1]

Hairy ghost pipefish

Solenostomus snuffleupagus

- Officially described as a new species in 2026, the hairy ghost pipefish is a tiny, highly camouflaged fish native to the southwestern Pacific.
- First spotted by marine biologist Dr David Harasti in Papua New Guinea in 2001, it took over two decades and several expeditions to find, collect, and formally describe a male-female pair.
- Because of its hairy snout and body, they reached out to the *Sesame Workshop* for approval to name it *Solenostomus snuffleupagus*; it is an evolutionary marvel hiding in plain sight.
- Despite its delicate, elegant appearance, this little animal is a predator. Micro-CT scans of its gut contents revealed whole skeletons of tiny fish. This overturned the long-held assumption that ghost pipefish survived exclusively on tiny crustaceans.
- Growing to just about 3 to 5cm in length (roughly the size of an AirPod), its body is covered in long filaments. It floats nearly motionless, swaying back and forth in the water to perfectly mimic floating red algae.
- Genetic analysis and CT scans have shown that this species diverged from its closest cousin about 18 million years ago. It also boasts 36 vertebrae—more than any other known ghost pipefish.
- Elusive Nature: Much like its elusive Sesame Street namesake, the species was incredibly difficult for scientists to track down.
- As a close cousin of seahorses and sea dragons, this fish reverses typical gender roles. Like other ghost pipefish, they have a unique reproductive cycle compared to seahorses. The females are larger than the males and have specialised, enlarged pelvic fins to form a brood pouch to carry up to 350 eggs.
- This elusive fish resides in the tropical waters of the southwestern Pacific, ranging from Australia's Great Barrier Reef eastward to Papua New Guinea, Tonga, and Fiji.

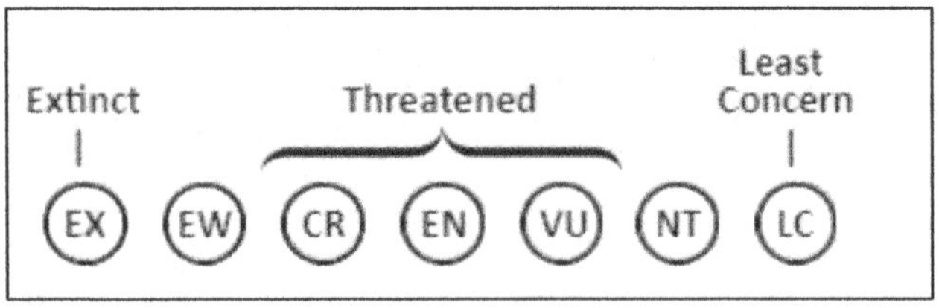

[1]

*Only officially described as a new species in 2026, it could be some time before this pipefish makes it onto the IUCN Red List!

Green Turtle

Chelonia mydas

- Even though green turtles are very large as adults, they are only a tiny two inches when they are born!
- Green turtles spend most of their time underwater and can be there for up to five hours before coming up to breathe.
- Green sea turtles are among the world's largest turtles, weighing between 65 and 130kg and measuring between 1 and 1.2m long. Their strong, teardrop-shaped shell covers most of their body except for their head and four flippers.
- Why are they called "green" sea turtles? Because it is thought that their unusual coloured fat is a result of their veggie diet; the green turtles eat **marine** plants such as seaweed and seagrass unlike most other sea turtles.[9] They used to be **extensively harvested** for turtle soup but now are protected.
- These turtles nest along the coastlines of more than 80 countries worldwide, with the largest nesting populations in Costa Rica and Australia.
- Green turtles **migrate** long distances between their feeding grounds and mating/nesting sites, with recorded distances over 2,600 kilometres. Females usually travel thousands of miles from their feeding grounds back to the beaches they were born on every two to four years, whereas males may make the journey **annually**.
- As powerful swimmers, turtles propel through the water with strong paddle-like flippers. Although they prefer to cruise along at around 3km/h, if threatened, they can reach speeds of up to 35km/h.
- After mating in the shallow waters near the shore, the female comes ashore to a sandy beach, where she uses her flippers to dig a nest. She lays a **clutch** of approximately 115 eggs, covers them with sand, and then returns to the ocean.[9] The temperature at which the eggs **incubate** influences the sex of the **hatchlings**. Below 27.7°C, the **hatchlings** will be male, above 31°C, they will be female, and between these temperatures, the **broods** will hatch both sexes.
- After about two months, the babies will use a special "egg tooth" to break their shells and hatch from their eggs.[9] They dig themselves out and must cross the sand to get to the water, avoiding many hungry **predators** such as birds, crabs, wild dogs, and lizards! A very dangerous journey.
- During the early years of a green sea turtle's life, they drift in the open ocean, feeding on **plankton**. As they mature, they **migrate** to coastal areas, including bays and lagoons, where they **graze** on seagrass.[9]

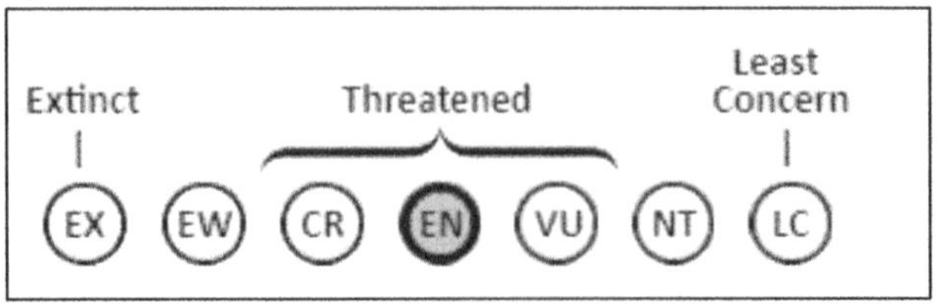

[1]

NCA List = V = Vulnerable

EPBC List = V = Vulnerable

GLOSSARY – All those big or unusual words! ...

Some of these words are really unfamiliar, even to me as the author of this book, but I have included them because I think they could be useful to anyone, child or adult, who is interested in nature. For the most part, a synonym (a word that means the same thing) has been used to try to keep these explanations as simple as possible. Definitions and synonyms from Dictionary.com and Thesaurus.com have been used on occasion. These definitions are not intended to be 100% perfect, but to give younger readers (and perhaps the adults reading with them) additional ideas to help them understand and explain unfamiliar words in the context of this book! Not all terms will be found in the text; however, they are still good terms to learn for any enthusiastic nature lover.

---

abdomen: belly, gut, midsection
aboriginal: original, earliest, native
aboriginal mythology: The "Dreaming" - the rich oral traditions, stories, and beliefs of indigenous Australians, passed down through generations
absorb: take in, soak up
abundant: ample, plenty of, extensive
abyssal (ocean): relating to or denoting the depths or bed of the ocean, especially between about 3000 and 6000m
abyssopelagic (ocean zone): abyssal zone
accumulate: to gather or collect something over time, causing it to gradually increase in quantity
acid, acidic: substance like lemon, sour
acoustic: relates to sound, hearing, or the science of sound
acrobat, acrobatic/s: like gymnastics, athletic
acronurus: highly specialised, transparent **larval** stage of surgeonfishes, tangs, and unicornfishes
adapt, adapted: change to suit, alter
adaptation/s: adjustment/s, alteration
adept: highly skilled or proficient at something
adhere, adhesive: stick, sticky like glue
adjoin/s, adjoining: be next to and joined with
aggregation/s: gathering many individual pieces and grouping them together
aggressive: ready or likely to attack or confront
agile, agility: athletic, quick, energetic
agriculture, agricultural: the science or practice of farming
akin to: similar nature or character; related by blood
albino, albinism: pale or white colouring, usually with pink eyes
alga, algae: plants living in water
altitude, altitudes: height in relation to sea or ground level
ambush: trick, trap, a surprise attack
ambush hunter, ambush predator: an animal that prefers to capture or trap prey by stealth or strategy
amphibian: an animal that can live both on land and in water, a cold-blooded vertebrate animal
amphipod/s: small, shrimp-like crustacean known for its flattened body
ampullae of Lorenzini: specialised sensory organs that act as a "sixth sense" for sharks and rays
anal: related to or near the anus
analysis: study, test, investigate
anatomy/anatomical: structure of a plant or animal, the way it is built
ancestor: an earlier species in biology
angler: an animal/person who catches fish
annual, annually: yearly, once a year
antenna, antennae: feelers
antennules: small antenna, especially one of the foremost pair of a crustacean
anterior (anatomy): towards the front

antivenom: a substance used to neutralise or counteract the effects of a venom
anus: the opening at the end of the digestive system
apex: the top, highest point, or pointed tip of something
apex predator: animal at the very top of a food chain or web that has no natural predators of its own
appendage/s: a part that is attached to a larger body
apex predator: a meat eating animal that preys on other animals but has no natural predators of its own, at the top of the food chain
aquarium, aquaria: fishbowl, fish tank
aquatic: occurring in water
aquatic vegetation: plants that rely on wet environments
arboreal: living in or among trees
archaea: microorganisms which are similar to **bacteria** in size and simplicity of structure but radically different in molecular organisation. They are now believed to constitute an ancient group which is intermediate between the **bacteria** and **eukaryotes**
arthropod: an animal with a segmented body, jointed limbs and usually a shell including insects, spiders, crustaceans
artificial: not natural, made by humans
ascidian: a type of marine invertebrate more commonly known as a sea squirt. They are sac-like, rubbery creatures that permanently attach themselves to rocks, shells, or ship hulls on the ocean floor.
asexual/asexually: independent of sexual processes
assimilate: ingest/digest, soak up
asymmetry, asymmetrical: state where two halves or sides of something do not match
athletic: physically strong, fit, and active
Austral spring/summer: a season in the Southern hemisphere
axis: an imaginary line that goes through the middle of something
bacteria, bacterium: a member of a large group of unicellular microorganisms which have cell walls but lack organelles and an organised nucleus, including some that can cause disease.
bacterivore, bacterivorous: an organism which obtains energy and nutrients primarily or entirely from the consumption of bacteria.
baitfish: any small fish that acts as a food source for larger, predatory fish
baleen: stiff, bristly material found in the mouths of certain whales instead of teeth; it works like a giant sieve or comb
baleen whale: giant ocean mammal that has stiff, comb-like plates in its mouth instead of teeth
barb: a sharp projection near the end of an arrow, fish hook, or similar object, which is angled away from the main point so as to make extraction difficult; a fleshy filament at the mouth of some fish such as barbel and catfish
barbed: having a barb or barbs
barbel: a fleshy filament growing at the mouth of some fish
bask: to lie or relax in pleasant warmth
bathypelagic (ocean zone): midnight zone
benthic: relating to or occurring on the bottom of a body of water
bill (of birds and some fish): beak; the hard, pointy part of a bird's mouth
biocide/s: a substance that kills life
biodiversity: differences in biology within plant and animal species
bioerosion: the wearing away or breakdown of hard natural materials—like rock, coral, or shells—by living organisms
biology: the scientific study of life and living organisms

biomass: any renewable organic material derived from plants and animals that is used to create energy
bioregion: a geographical area defined by its natural ecological systems and physical features—such as mountain ranges, soil types, and watersheds—rather than by artificial political border
bite force: force applied between the upper and lower mouth structures, jaw and/or teeth
bivalve: any mollusc with two shells hinged together
botany: the scientific study of the physiology, structure, genetics, ecology, distribution, classification, and economic importance of plants
brackish: water that is saltier than fresh water but not as salty as seawater
broadcast spawner: A broadcast spawner is an aquatic animal that reproduces by releasing massive amounts of eggs and sperm directly into the water, relying on the currents for fertilisation
brood: a number of offspring, clutch, youngsters
brood pouch: a pouch where eggs develop and hatch
bryozoan: (often called a "moss animal") is a tiny aquatic invertebrate that lives in colonies
buoyant, buoyancy: the upward force exerted by a fluid (like water or air) that opposes the weight of an object immersed in it, making objects float or rise
burrow: a tunnel or hole that an animal digs as a place to live or hide; the act of digging down into sand or mud
bycatch: the catch of unintended marine species or sizes of wildlife during commercial or recreational fishing
calcium carbonate: a common chemical compound found abundantly in nature. It is the main ingredient in rocks like limestone, chalk, and marble, and makes up the shells of marine animals and eggshells
camouflage: disguise, mask, concealment
canine/s: a pointed tooth between the incisors and premolars of a mammal, often greatly enlarged in carnivores
cannibal, cannibalism: any animal that eats the flesh of its own kind
captive breeding: bred in captivity
captivity: kept in a place like a zoo or a cage
carapace: a bony or chitinous shell, test or shield covering some or all of the dorsal part of an animal, e.g. turtle
carnivore, carnivorous: meat-eating
carpal: relating to the bones forming the wrist
carrion: dead and rotting flesh of an animal
cartilage: strong, bendy, and rubbery tissue
cartilaginous tissue: made out of cartilage like our ears and nose tips
caudal (anatomy): of, relating to, or being a tail; directed toward or situated in or near the tail or posterior part of the body
cavity: an unfilled space within a mass (especially a hollowed-out space)
cay: small, low-elevation, sandy island formed on the surface of a coral reef
cephalofoil: the technical name for the unusual, flattened, and wing-like head of a hammerhead shark
cephalopod: an advanced marine animal—such as an octopus, squid, cuttlefish, or nautilus. The word literally translates from Greek to mean "head-foot," describing how their arms and tentacles are attached directly to their head
ceremony: a formal event with a specific purpose
cetacean: marine mammals belonging to the order *Cetacea*, which includes all species of whales, dolphins, and porpoises. They are fully adapted to life in the water but are warm-blooded, breathe air through a blowhole, give birth to live young, and produce milk
characteristic: distinctive, singular feature
chemical cues: hints that are chemical, could be

particular smell
chevron: simple geometric shape that looks like a "V"
chromatophore: tiny colour-changing cells under an animal's skin
ciguatera: type of food poisoning caused by eating reef fish contaminated with
ciguatoxin: a potent, tasteless neurotoxin produced by marine algae (dinoflagellates) that grow on coral reefs
circulate: move in a circle, pass from place to place, spread
circumference: the distance/measurement all the way around an object/circle
cite, cited: quote, name, reference
clade: a fundamental grouping in evolutionary biology consisting of a single common ancestor and *all* of its descendants
claspers: modified pelvic fins
classify, classification (scientific): the sorting of things into groups based on how they are alike
clime/s: the average weather conditions of a particular area
cloaca, cloacal: a shared exit for bodily waste and reproductive function
clutch (of eggs):total number of eggs/young in one nesting
coastal: anything located, found, or happening near a coast
coastal dredging: excavating material from a coastal water environment
cocoon: a protective, silky case
co-exist: live or exist together
colony, colonies, colonisation: group, settlement, swarm
colour morph: a specific body colouration variation
combatant/s: fighter, battler
commensal: an organism that lives in a close relationship with another species (the host), where it benefits by getting food, s helter, or support, while the host is neither helped nor harmed
commerce: business, trade
commercial: merchant, business, of trade, public
communal: shared or used by all members of a group
compact: something is closely packed or a small size, taking up little space
complex: complicated, convoluted
compress, compresses: to press or squeeze something so it takes up less space, time, or size
condense: to make something shorter, more compact, or more concentrated
cones (eyes): specialised cells that allow us to see colour and fine details, particularly in bright light
configuration: composition, shape, structure
congregate: gather into a crowd or mass
conserve, conservation: saving and protecting
conspicuous: easily seen or noticed; standing out of a group
continental island:
constrict, constriction: to become tighter or narrower, or to make something become tighter or narrower
construct: build or put together
consume: eat, drink or use something up
contamination: infection, pollution, corruption
continent: big, solid area of land – Earth has seven continents
conspicuous: very easy to see, notice, or attract attention
co-operative: working with others to help make things happen
copepod: a tiny, shrimp-like crustacean with "oar feet" that lives in nearly all aquatic environments
copious: abundant, plentiful, or large in quantity or number
coral cay: a small, low-lying island formed on the surface of a coral reef. It is created when ocean waves and currents deposit sand, coral fragments, and other debris onto the reef, which gradually builds up above sea level to form an island
corallite: the tiny, hard, cup-shaped skeleton built by an individual stony coral polyp

corallivore: an animal that eats coral
coronal (anatomy): dividing body into front and back
counterpart: equals, fellow, like, twin
courting, courtship: set of display behaviours in which an animal, usually a male, attempts to attract a mate
covet, coveted: strongly desire or want something
cranial (anatomy): towards the head
creature: a living animal or living insect
creche: a group of young cared for communally
credible: believable, reasonable, likely
creek: small stream or river
crepuscular: resembling or relating to twilight
crevice: slit, crack, cleft, gap
critical: very important
critically endangered: facing an extremely high risk of extinction in the wild
critter: informal word for any living creature
crucial: extremely important or necessary
crustacean/s: aquatic arthropod typically with body covered with a hard shell or crust, such as lobsters, shrimps, crabs, barnacles
cryptic: secret, mysterious, hidden
ctenoid scales: scales with spines or comb-like teeth along their free edge
culprits: wrongdoer, offenders
cultivate: breed, raise, tend to, plant
cultural significance: of some value for past, present or future generations
curious/curiosity: having a strong desire to learn, explore, or understand something
currency: money, legal tender
cyanobacteria: formerly known as blue-green algae, are photosynthetic microscopic organisms that are technically bacteria
cycloid scales: thin, smooth and circular or oval-shaped fish scales found on many soft-finned bony fish
cylinder, cylindrical: 3D shape like a tube with circles on either end
debris: scattered pieces of rubbish or remains
decade: 10 years
decline, declining: drop, gall, decay, weakening
decurved: curved downward or bent downward
deep (anatomy): further from the surface
defence mechanism/s: adaptations that help avoid predators or mitigate adverse effects.
defence, defensive: protecting or defending against something
degrade, degradation: reduce in quality, lessen, weakening, deterioration
demersal: living or found near the bottom of a body of water
den: the lair of a wild animal
dense: thick, heavy, impenetrable, solid
derivation (words): where the word/s came from
dermal: of, relating to, or located on the skin
dermal denticle: the tough, V-shaped "skin teeth" that cover the bodies of sharks and rays
dermal spinules: a microscopic, thorn-like, or spine-like projection found on the skin of certain animals
descendant: a person, animal, or plant that comes directly from a specific ancestor or earlier generation
destruction: breaking something apart, causing so much damage it can't be used or fixed, or even removed
deter: to discourage or prevent someone or something from doing something, usually by making them afraid of the consequences or aware of the difficulties
detritus: small pieces of disintegrated material such as organic matter
diagnosis, diagnostic: the process of identifying a disease, condition, or injury by looking at its signs and symptoms
difference between endemic and native? An endemic species is a native species but found only in a particular area, large or small

differentiate: to recognise, state, or show the difference between two or more things
digest: to break down food into smaller pieces to use for energy and growth
digestive system: the parts of the body that work together to turn food and liquids into the building blocks and fuel that the body needs for energy and growth
digit/s: finger, thumb or toe
dimorphic, dimorphism: two different forms
dioecious: having the male and female reproductive organs in separate individuals
disease: illness or sickness characterised by specific signs or symptoms
disguise: to give a new appearance to a thing, especially in order to hide its true form
disorient, disoriented: cause to lose sense of direction, feel lost, or become mentally confused
disperse/dispersal: to spread out or move in different directions
dissipate: disappear, waste or fritter away
distal (anatomy): further from the centre
distinction/s: noticing the difference between two things
distinguish, distinguishes: mark off as different
distinctive, distinctively: having a special quality, style, colour, etcetera
distribute, distribution: divide into, disperse through a space or over an area, spread, scatter
diurnal: of or during the day
diverse, diversity: different or having variety
dominant, dominance: more strong than others
dormant: inactive or not active, but capable of becoming active later
dorsal, dorsum (anatomy): of, relating to, or situated at the back, or dorsum of
drainage: removing excess water from an area
drey: the nest of a possum or certain other animals, like a squirrel
drought: a long and abnormally dry spell
dune: mound, hill, or ridge of sand piled up by the wind or water
dusk: time of day when the sun is setting and it's starting to get dark, just before night
dusky (of colour): of a darker colour, shadowy
dwindle: become less or smaller gradually
echolocation, echolocate: the location of objects by reflected sound
ecology: the scientific study of how living organisms interact with each other and with their physical environment
ecosystem: a geographic area where plants, animals, and other organisms, as well as weather and landscapes, work together to form a bubble of life
edible: fit to be eaten as food, eatable
eel: long snake-like fish
elaborate: complicated, detailed, **complex**
electroreceptor/s: specialised sensory organ/s that allows certain animals to detect weak electrical fields in their environment
elusive: tricky, difficult to catch, shy
endemic: native species found only in a particular area
embryo: baby in its very early stages, before it's born; animal or plant that's just starting to grow and develop
embed/embedded: firmly fix an object deeply into a surrounding substance, or to integrate content
endangered: there is a very high risk of extinction due to rapid population declines of 50 to more than 70% over the previous 10 years, a current population size of fewer than 250 individuals
environment, environmental: the natural surroundings that make up the world we live in
EPBC: Environment Protection and Biodiversity Conservation Act 1999 (EPBC Act) is Australia's main environmental law
ephemeral: lasting for a very short time; fleeting or transitory

epic: very large, grand, or impressive
epipelagic (ocean zone): sunlight zone
estimate, estimated, estimating: guess, projection
estuary, estuarine: an area where a freshwater river or stream meets the ocean
eukaryotes: any cell or organism that possesses a clearly defined **nucleus**
European settlement: 1788-1810 early European settlement
evade: avoid or get away from something, often by being clever or sneaky
evolve, evolved: grow, mature, advance
evolution, evolutionary: change, expand, progress, transform
excavate: dig out, burrow, uncover
excrete: to pass or discharge waste materials from a living body
excretion/s: discharge by excreting
exotic: of foreign origin or character, introduced from abroad
extensive, extensively: covering a large area
extinct, extinction: having died out, no longer existing
extreme weather events: occurrences of unusually severe weather or climate conditions that can cause devastating impacts on communities and agricultural and natural ecosystems
exudate: substance coming out, ooze
fatal, fatalities: causing death
fatigue: an overwhelming, persistent sense of exhaustion and lack of energy
fauna: all the animal life in a specific region or time period
faunal region: any of six or seven areas of the world defined by animal geographers on the basis of their distinctive animal life
fertilised, fertilisation: make productive, capable of development
filament, filamentous: a single, thin, thread-like object or structure
fission-fusion society: an animal social structure where the size and composition of a group change dynamically over time
flamboyant: marked by or given to strikingly elaborate or colourful display or behaviour that attracts attention flank/s: side between ribs and hips
flock: group of animals, especially birds like geese or chickens, or sheep and goats, that stay together
flora: all the plant life in a particular region, time, or environment
fluke: either of the two flat, horizontal lobes that make up a whale's tail
foliage: the leaves of a plant or tree collectively, or leaves in general
folivore: an animal or organism that primarily eats leaves
follicle/s (biology): small, sac-like cavity or pouch in the body that holds or grows something
forage, foraging: look for food
formidable: something that is really strong, impressive, or even a little scary, but also something you might respect
fossil: the preserved remains or traces of a living thing that lived long ago
fragile: something that is easily broken, damaged, or destroyed
fragment, fragmented, fragmentation: broken off, scattered, isolated, separated
frugivore, frugivorous: an animal that thrives mostly on raw fruits or succulent fruit-like produce of plants such as roots, shoots, nuts and seeds
fundamental: basic, essential, or the most important part of something else
fungus, fungi, fungal: simple organism, or living thing, that is neither a plant nor an animal
gastrointestinal: the organs of the body that make up the digestive tract, specifically the stomach and intestines
gelatinous: having a thick, sticky consistency that resembles jelly

generalist feeder: feeds on a wide variety of things and thrive in various environments

generic: referring to all members of a genus

gene/s: the basic unit of heredity. This means that genes determine what traits are passed down from a female and male to their offspring. Genes exist inside the cells that make up living things

genetic/s: traits carried in genes, which are like special instructions for making a body

genus: the major division of a family, usually of more than one species

geology, geological: the scientific study of Earth, including its physical features, its history, and the processes that have shaped it, such as rock formation and plate tectonics

gestation: the process or period of developing inside the womb between conception and birth

gill: an organ used by aquatic animals to breathe by extracting oxygen from water and releasing carbon dioxide

gizzard/s: a muscular organ in the digestive tract of some animals that grinds up food

gland/s: an organ in the human or animal body which secretes particular chemical substances for use in the body or for discharge into the surroundings

global/globally: worldwide, comprehensive

globose: having the shape of a globe or sphere; round

glossy: shiny, lustrous, polished

gorgonian: a type of soft, colonial marine animal that includes sea fans and sea whips; an octocoral

graffiti: writing or drawings, scribbled,scratched, or sprayed, usually illicitly

granular: something that is made of tiny grains or particles, like sand or sugar, or that looks like or feels like it is made of those small bits

graze, grazing: eating, feeding, nibbling, munching

gregarious: sociable, outgoing, and fond of being a round other people

habitat/s: place where usually found, natural environment of an organism

harass: to repeatedly trouble, annoy, or bother someone in a way that makes them feel upset, threatened, or uncomfortable

harvest, harvested: gather food

hatch, hatching: baby animal coming out of its egg, or that an egg is breaking open to release the baby

hatchling: young bird, reptile or fish just hatched from egg

haven: a place of safety and refuge

hectocotylus: a specialised arm on male cephalopods (like octopuses) that is adapted for reproduction

herbicide/s: a substance that is **toxic** to plants

herbivore/herbivorous: an animal that primarily eats plants

herd/herding(prey): the strategic manipulation and control of animals by predators

hermaphrodite/s: an organism that naturally possesses both male and female reproductive organs, allowing it to produce both sperm and eggs

heterosexual: attraction to the opposite sex

hibernate, hibernation: spend the winter in a dormant state

hinder: make harder or slower to do, make difficult

honeydew: sugary material secreted by some insects such as aphids, scale

horizontal: parallel to the horizon

host: an organism that provides shelter and nourishment to another organism, which can be a parasite, virus, or microbe

hostile: aggressive, warlike

human disturbance: people disturbing the environment: for example - tree removal. road building construction near shorelines and streambanks

humid, humidity: moisture or water vapour in air

hydrodynamic: the branch of physics that studies how liquids (like water) move and how forces act upon them

hydrology: the branch of science concerned with the properties of the earth's water, and especially its movement in relation to land.

iconic: representative of

illegal: against the law, not legal

imminent: likely to happen very soon

imperilled: put in danger or at risk

immature: something that is not fully developed or grown

immune to: not being affected by something

impress, impressive: to cause to feel admiration or respect, often by doing or showing something exceptional

incisor/s: a narrow-edged tooth at the front of the mouth, adapted for cutting

incubate, incubated, incubation: brooding

indigenous: having originated in and being produced, growing, living, or occurring naturally in a particular region or environment

indigenous people: First Nations people

inedible: not fit or safe to eat

inferior (anatomy): below, lower than

infertile: unable to reproduce naturally

infiltrate, infiltration: the process of cells or substances moving across a barrier, typically a tissue barrier, into a place they are not normally found, or in which they are typically found in lower concentrations

influx: the arrival or entry of many things

ingenious: exceptionally clever, inventive, or resourceful

ingest/ed: to take something in usually by eating, drinking, or swallowing

inhabit, inhabits: live in, occupy, populate, reside

innate: a quality or ability that exists from birth rather than being learned

interlock: to connect or join two or more things so that they are mutually dependent

insatiable: impossible to satisfy

insecticide: an agent (usually chemical) that destroys insects

intense: extreme, very strong, existing at a high degree

intertidal: between high and low water marks

intimidate/d: scare, frighten, bully, terrify

introduce/d (of species): plant or animal that wasn't originally there but was brought in by humans, either on purpose or by accident

inundate: overwhelm or flood

invasive: threatening a plant or animal community, by spreading prolifically

invertebrate/s: animals without backbones

iridescent: something that changes colours when you look at it from different angles

iridophores: an iridescent chromatophore

iris (of eye): the coloured tissue at the front of the eye that contains the pupil in the centre

isolation, isolated, isolate: detach, separate, place apart

IUCN: International Union for the Conservation of Nature

juvenile/s: youth, junior, infant, young

keel (fish): a hard, rigid ridge running along the body, most commonly found on the belly or just ahead of the tail

keystone species: an organism that has a disproportionately large impact on its ecosystem relative to its abundance

kinetoplast, kinetoplastids: a group of flagellated protozoans that are distinguished by the presence of a DNA-containing region, known as a "**kinetoplast**", in their single large mitochondrion

krill: tiny, translucent, shrimp-like crustaceans

laceration: deep cut or tear

lactate, lactating, lactation: producing and secreting milk from the mammary glands

lagoon: a shallow body of water, usually saltwater or brackish, separated from a larger body of water like the ocean by a natural barrier such as a sandbar, barrier island, or coral reef

lamina (biology/botany): a thin layer or plate

larva, larvae, larval: immature stage that undergoes complete metamorphosis

lateral (anatomy): towards the side

leverage: using what you already have to get a bigger advantage

life cycle: the series of changes in the life of an organism

lifespan: the period of time of existence from birth to death or the end of useful function

litter: number of young born in one birth

littoral: relating to or situated on the shore of the sea or a lake

littoral rainforest: a closed forest community that grows near the ocean, usually within 2km

lobe, lobed: curved, rounded, or clearly defined section

local, locally, localised: restricted to one place

locally extinct: no longer present in a specific geographic region but still exists elsewhere

locomotion: the act of moving from one place to another

longitudinal, longitudinally: running lengthwise rather than across.

lore: the region between the eyes and nostrils of birds, reptiles, and amphibians

lure: attract, tempt, or trick by offering a reward

macro- (prefix): large; large-scale.

macroalga/e: a large, multicellular form of algae that is visible to the naked eye, commonly known as seaweed

mammal: warm-blooded vertebrate having hair or fur, producing milk for the young and giving birth to live young – exceptions are the **monotremes.**

manoeuverable, manoueverability: easily moved, steered, or controlled into different position

mantle (biology): a soft, fleshy outer covering of molluscs and brachiopods that is the body wall of the animal

Māori people: indigenous Polynesian people of mainland New Zealand

margin: boundary area

marine: of or relating to the sea

marsh, marshes, marshland: type of wetland ecosystem characterised by poorly drained mineral soils and by plant life dominated by grasses

maternity: the period during pregnancy and shortly after childbirth

matriarch, matriarchal: a social system in which positions of power and privilege are held by females

mature, maturity: fully developed, having completed natural growth and development

mechanism: the fundamental processes involved in or responsible for an action, reaction, or other natural phenomenon

medial (anatomy): towards the middle

melody, melodious: sounding pretty and musical

membrane/s: sheath, sheet, film of tissue

mesial: being or located in the middle or a median part

mesmerise/mesmerising: fascinate, captivate, spellbind, enthrall

mesopelagic (ocean zone): twilight zone

metabolic, metabolism: the chemical reactions in cells that change food into energy

metamorphosis: change in life cycle e.g. pupa to caterpillar to butterfly

metapopulation: population of populations

methodical: orderly, structured, deliberate

micro- (prefix): very small, tiny

microalga/e: microscopic algae invisible to the naked eye; phytoplankton typically found in freshwater and marine systems, living in both the water column and sediment. They are unicellular species which exist individually, or in chains or groups

microendemic: species with very small geographic distributions (ranges)
microorganism: an organism that is so small it can only be viewed using a microscope. Also called microbes, these tiny living things are found everywhere in water, soil, air, and the human body, with examples including bacteria, fungi, and protozoa
microphyte: a microscopic plant
microscopic: incredibly small, unable to be seen without a microscope
migrant: drifter, traveller, wanderer
migrate, migration, migrating: going from one country, region or place to another
migratory: characterised by migration
milt: seminal fluid
mimic, mimicking, mimickry: copy, simulate, imitate
mitochondria, mitochondrium: the tiny powerhouses inside our cells! They break down food we eat to give us energy, so we can move, think, and do everything else we do - think of them as little energy factories
moist: slightly wet, damp
moisture: the presence of a liquid, especially water, often in small amounts
molar: the back teeth in the mouth, shaped like flat, grinding surfaces
mollusc/s: animal that has a soft body, no spine, and is often covered with a shell
monitor: observe, watch, keep an eye on
monoecious: having both the male and female reproductive organs in the same individual
monogamous: having only one mate at a time
monopolise: have complete and sole control over, take over control
morph (biology): one of the distinct forms or variations found within a species, especially when these variations are visually or behaviourally different
morphology, morphological: form of a thing
motivate. motivated: inspire, persuade, prompt, spark
moult: shed, cast, peel - when an animal sheds its old outer covering (skin/feathers, for example) to make way for a new one or new ones
mucous: slippery, gel-like fluid
murky: dark, cloudy, difficult to see through
mutate, mutation: change, develop a completely new physical characteristic or form because of a permanent change in genes
mysid shrimp: small, shrimp-like crustaceans
myth, mythology: belief, legend, lore, tradition
native: an animal or plant that has lived in a specific area for a long time, naturally, without being brought in from somewhere else
navigate, navigation: find a route to a destination
navigator: a person or animal who navigates
NCA: Nature Conservation Act 1992
nectar: sweet, sticky liquid made by flowers
nectarivore, nectarivorous: an animal which derives its energy and nutrient requirements from a diet consisting mainly or exclusively of the sugar-rich nectar produced by flowering plants
neurological: relating to the nervous system
neurotoxin: poisonous substance that attacks the nervous system
niche: a specialised area of interest, a position that is a perfect fit for someone, or a specific role an organism plays in its environment
nocturnal: active at night, relating to night
nomadic: wandering, travelling, roving
no-take: not allowed to be taken/caught
nucleus, nuclei: the structure in a cell that contains the chromosomes
nudibranch: soft-bodied, shell-less marine sea slug
numerous: large number, many
nurture: care for, protect
nutrient/s: source of nourishment, food

nutrient-rich: rich in nutrients, nourishing food
oceanic: marine, of the oceans
ocelli: simple eyes found on many insects and other invertebrates that detect light, but do not form images. They function as sensors for ambient light intensity, help with flight stabilisation, and are involved in orientation and movement detection
ocelli (2nd meaning): eye-like spots or markings
oesophagus: a muscular tube in the digestive system that carries food and liquids from the pharynx (throat) to the stomach
omnivore, omnivorous: eating both animal and plant foods
ooze: soft, wet, or slimy mud; ocean ooze = fossilised remains of microscopic sea creatures
opportunistic: taking opportunities when they present
opposable, opposing: able to be placed opposite to something else
organelle: a **subcellular** structure that has one or more specific jobs to perform in the cell, much like an organ does in the body. Among the more important cell **organelles** are the **nuclei**, which store **genetic** information; **mitochondria**, which produce chemical energy; and **ribosomes**, which assemble proteins
organic: live, living, natural
organic material: matter that is made of organic compounds from the remains of organisms such as plants and animals, and their waste products
organism: animal, plant, structure, creature
ornamental: purely for decoration
ornithology: study of birds
ovary: the organ in a female animal that produces eggs and certain hormones; the part of a flower that contains the seeds and grows into a fruit
overexploit, overexploitation: overuse, excessive use of resources
overfish, overfished, overfishing: catching too many fish at once so that the breeding population is too reduced to be able to recover
overhunt, overhunted, overhunting: the act of hunting animals or birds to an excessive degree, which can be harmful to the population or the environment
overgraze, overgrazed, overgrazing: allowing grazing animals to reduce grazing land to beyond the point of recovery
oviparous: egg-laying
ovoviviparous: describes animals that produce eggs that hatch *inside* the mother's body
ovum: an egg cell
ovule: an outgrowth of the ovary of a seed plant that after fertilisation develops into a seed
palatable: pleasant tasting
paralysis: inability to move, can sometimes affect entire body systems such as breathing, movement, heartbeat
parasite, parasitic: an organism that lives on or in an organism of another species (the host) from the body of which it obtains nutriment
pardaxin: a toxic **peptide**
parthenogenesis: a form of asexual reproduction where an embryo develops from an unfertilised egg
pathogen: an organism or agent that causes disease, also known as an infectious agent or germ
pectoral fin: see fish anatomy diagram
peduncle (botany): the stalk of a flower
peduncle (ichthyology – the study of fish): the narrow part of a fish's body just before the tail fin

pelagic: occurring in the open sea
pelvic: having to do with the pelvis
pelvices: plural of pelvis
pelvic fin: see fish anatomy diagram
pelvis: the bony structure of the body below the abdomen that is located between the hip bones and contains the bladder and rectum
peptide/s: short chains of amino acids, which are the fundamental building blocks of proteins
perimeter/s: boundary, edge, borderline
perpendicular: two lines, planes, or surfaces meet at a perfect 90° right angle, forming an "L" or "T" shape
pesticide/s: a substance used for destroying insects or other organisms harmful to cultivated plants or to animals
pharynx, pharyngeal: muscular, funnel-shaped tube that connects the mouth and nasal cavity to the oesophagus and windpipe (larynx)
phytoplankton: consisting of microscopic plants
pigment, pigmentation: substance that gives colour to another material
piscivore/piscivorous: a carnivorous animal that primarily eats fish
placid: gentle, peaceful, quiet
planktivorous: feeding primarily on **plankton**
plankton, planktonic: floating, drifting, or somewhat motile organisms occurring in a body of water, primarily comprising microscopic algae and protozoa
plastron: the ventral part of a turtle shell
plateau: a large, flat, elevated area of land that rises sharply above the surrounding terrain on at least one side
plume/plumage: feather (singular), feathers of a bird, collectively
poach/poaching: to illegally hunt or fish animals
poison, poisonous: likely to cause serious harm or death
pollen: a fine powdery substance, typically yellow, consisting of microscopic grain discharged from the male part of a flower or from a male cone. Each grain contains a male gamete that can fertilise the female ovule, to which pollen is transported by the wind, insects, or other animals
pollination: the act of transferring pollen grains from the male anther of a flower to the female stigma to help make new plants
pollinator, pollinators: animal, insect or other agent that transfers pollen to a plant allowing fertilisation
pollutant: substance, chemical, or energy that causes harm, disrupts natural ecosystems, or makes natural resources unsafe
pollution: the introduction of harmful materials into the environment
polychaete worm: poly means 'many', chaete means 'hair', so these are a species of marine worm, commonly known as bristleworms.
polyclad, polyclads: any free-swimming, marine flatworm of the order Polycladida, having a broad, flat body and a many-branched gastrovascular cavity
polygynous: having more than one mate at the same time
polyoestrous: can raise several litters a year
polyp (zoology): small, tube-shaped animal with a mouth and tentacles at one end, and an anchor point for attaching to a surface at the other
population: number of inhabitants of an area
pore/s: any tiny opening on the outer surface of an animal that allows for the passage of fluids, gases, or wastes
posterior (anatomy): behind, towards the back
posture, posturing: behaviour that is intended to attract attention and interest
potent: having great power, strong influence, or a highly effective physical or chemical effect

precarious: dependent on circumstance, unsure, indeterminate, unreliable, doubtful

precipitation: what happens when water in the sky (clouds) falls back down to Earth as rain, snow, sleet, or hail

precision: the quality of being exact, consistent, and highly detailed

precocial: (of a young bird or other animal) hatched or born in an advanced state and able to feed itself and move independently almost immediately

predator/s, predation: an organism that exists by preying on other organisms

predict, prediction: to be able to tell in advance

pregnant, pregnancy: containing a developing embryo, foetus, or unborn offspring within the body

prehensile: adapted for seizing, grasping, or taking hold of something

premature: happening or done too soon, or before the expected or natural time

prey, preying: an animal, insect or organism hunted or taken for food, especially by a carnivorous animal

primary: most important, main

primitive (biology): an organism, cell, or trait that closely resembles an ancient ancestor and hasn't evolved much

prokaryotes: organisms whose cells lack a **nucleus** and other **organelles**. **Prokaryotes** are divided into two distinct groups: the **bacteria** and the **archaea**, which scientists believe have unique evolutionary lineages. Most **prokaryotes** are small, single-celled organisms that have a relatively simple structure

probe: a tool or an investigation used to explore or gather information about something that is hidden, unknown, or hard to reach

prolific: producing a great amount, high volume, or frequent results

prominent: standing out, obvious

protandrous: maturing as male first

protector/s: anything that keeps something safe from harm, damage, or danger

protogynous: organism that is born female but h as the ability to change into a male later in life

protozoa, protozoan: any of a diverse group of eukaryotes that are primarily unicellular, existing singly or aggregating in colonies, are usually non-photosynthetic and often classified further into phyla according to their capacity for means of motility

protuberance/s: a rounded, outward projection or swelling on a surface

protrude, protrusion, protrusable: stick out, jut out, or project beyond a flat surface or edge

provoke/d: cause a specific reaction or emotion

prowl, prowling: move around an area quietly, slowly, and secretly

proximal (anatomy): closer to (root of the limb)

proximity: nearness or closeness

pseudo: false, fake, not genuine

pulmonate, pulmonates: having lungs or lunglike organs

pupil (eye): the black opening in the middle of the coloured part of the eye

pursue: chase, track, follow, trail, hunt

pursuit: chase, hunt, quest, run

quarry (prey): something being hunted or pursued, specifically prey

quarry, quarrying: an open excavation usually for obtaining building stone, slate, or limestone

radiate, radiating: lines or pathways spreading outward from a centre

radio-tracking: a way of sending sounds or other information through the air from a transmitter attached to the animal to enable researchers to track it

range (of species): an area where a member of a particular species can be found during its lifetime

rapid/rapidly: quickly, quick

raptor: usually a bird or prey, such as an eagle or a hawk

rare: not common or usual

recognise, recognition: seeing something you have seen before or can clearly identify by other means
recreation, recreational: fun activity, amusement, hobby
reflex: an automatic, involuntary, and nearly instantaneous physical response to a stimulus that does not require conscious thought
refuge: place that offers safety and protection from danger, trouble, or bad weather
regenerate, regeneration (biology): the process of renewal, restoration, and tissue growth that makes genomes, cells, organisms, and ecosystems resilient to natural fluctuations or events that cause disturbance or damage. Every species is capable of regeneration, from bacteria to humans.
regimen: a planned, systematic routine or set of rules followed consistently over time to achieve a specific goal
remnant: remains, fragment, piece
reproduction, reproduce, reproductive: the production of offspring
reptile: cold-blooded vertebrate
resident: permanent dweller
residential development: housing development
respiratory system: breathing system, lungs
restrict, restricted: confine, restrain, regulate, limit
reticulated: constructed, arranged, or marked like a net or a network.
retina, retinae *(Pl.)***:** inside back of eye of several layers
retract, retractable: drawing a part of a body back into itself
revere, revered: respect or admire greatly
reverse: backward, contrasting, opposite, contrary, change
reversible (physiological): something that can be changed from back to front or front to back
rhomboid, rhomboidal: parallelogram in which adjacent sides are not equal
rhythm, rhythmic: the timing of sounds, silences, and movements in a regular, repeated pattern
ribosome: an intercellular structure made of both RNA and protein, and it is the site of protein synthesis in the cell
rigid: not able to be bent or changed easily, stiff
ritual: a sequence of actions or words performed in the same order each time, often for a special purpose like a religious ceremony or a routine
rival: animal or plant that competes for the same goal or in the same area
river delta: fan- or triangular-shaped landform formed at the mouth of a river where it empties into a larger body of water
riverine: of, relating to, or resembling a river. It can also describe something located or living on or near the banks of a river
robust: strong and healthy
rods (eyes): specialised cells which are sensitive photoreceptors to light and help us see in low-light conditions. They are primarily responsible for night vision and peripheral vision, but they don't perceive colour
roost, roosting: perch where birds rest or sit
rostrum: beak, snout, or beak-like structure
rufous: tinged with red, brownish red
rump: hindquarters, buttocks, sacral region of a bird
rural: relating to the countryside and not to towns or cities
sagittal (anatomy): dividing body into right and left
saline: containing or impregnated with salt
salinity: saltiness
samara: a dry indehiscent usually one-seeded winged fruit
sanctuary, sanctuaries: a place of refuge or protection, offering safety from danger or hardship
satellite: an artificial body placed in orbit round the earth or moon or another planet in order to collect information or for communication.

satellite tracking: using satellites orbiting earth to track
satellite transmitter: device which sends a signal to a satellite
saturate, saturated: soaked, soggy, waterlogged
scarce, scarcity: not plentiful, hard to find or hard to get
scavenge/scavenger: an animal that feeds on the remains of other animals or eats decomposing plant or animal matter. This means that scavenger animals can be both herbivores and omnivores. Most times, scavengers do not kill their own prey, even if they may possess the ability to do so
school, schooling (of fish): group of fish moving together
scleractinian: hard or stony coral that builds its own calcium carbonate skeleton; these marine animals form the foundational framework of the world's coral reefs
sclerophyll forests : vegetation that has adapted to long periods of dry, hot weather
sea squirt: invertebrate, usually fixed potato-shaped organism using sac-like filter feeding
secretive: keeping secrets
sedentary: staying in one place, not moving much
sediment, sedimentation: organic matter settling to bottom of a liquid
sedimentary: of or relating to rocks formed by the deposition of sediment
segment/s: part of, divided parts
semi- (prefix): half, partially, incompletely
semi-parasitic: an organism which gets water or food from host but can also produce its own food
semi-prehensile: able to grasp but cannot really hang on tightly
semi-transparent: almost see through
sensory: relating to sensation or to the senses
serrated: having a jagged or toothed edge, like a saw
seta/setae (*Pl*): a stiff, bristle-like hair or process on an organism, with different meanings depending on the context. In biology, it can refer to sensory and locomotive hairs on invertebrates like spiders and earthworms, the stalk of a moss capsule, or a bristle on a caterpillar
settlement: a place, typically one which has previously been uninhabited, where people establish a community
sequential: following a specific order or happening in a clear, step-by-step sequence
shellfish: seafood like molluscs, crustaceans
sibling: each of two or more children or offspring having one or both parents in common; a brother or sister
sift: an act of sifting something, especially so as to isolate that which is most important
silhouette: a dark shape seen against a light surface
significant, significance: very important or has a special meaning
silt: fine sand, clay or other material carried by running water
siltation: water pollution by silt
simultaneous, simultaneously: happening, existing, or being done at exactly the same time
skim, skimming: to pass lightly or hastily; glide or skip along, above, or near a surface
snout: the long nose and mouth of an animal that sticks out
social, socialise: friendly, like company
solitary: by itself, single
sophisticated: complex and intricate
spawn, spawning: reproduce, generate (usually fish or frog eggs)
sparse: something that is thinly spread out or not dense
specialised/specialisation:
species: one of the classes of things included with other classes in a genus
specific: clear-cut, particular, precise, unique

specimen: sample, example
sperm: the male reproductive cell, also known as a gamete, which contains the father's genetic material and is responsible for fertilising a female egg to begin the development of a new individual
spinose: having many sharp, pointed things sticking out; spiny, prickly
spinule/s: very small spine, thorn, or prickle
spire: a part of the coiled shell of molluscs. The spire consists of all of the whorls except for the body whorl
spur (biology): an outgrowth of bone covered in a sheath of horn found in various anatomical locations in some animals
staccato: detached, short, and abruptly disconnected
stalk (hunt): walking quietly in search of animals or in pursuit of an individual animal
stalk: pursue without being seen or heard
stalling: stopping (flying)
stance: way of standing physically or figuratively
stealth: movement that is quiet and careful in order not to be seen or heard
strategy, strategic: plan designed to achieve a specific goal
strident: loud and harsh, or describes a forceful, insistent, and often irritating way of expression
stochastic: random or involving chance
storm front: border between two different kinds of air, often with different temperatures; when a front is created, weather such as storms and **precipitation** happen
streamers (tail feathers): tail streamers on birds are the elongated narrow tips of the tail feathers; elongated, specialised tail feathers that are often used for display and communication, particularly during mating rituals. They can be quite long, sometimes exceeding the bird's body length, and are often vibrant in colour or pattern
streamline, streamlined: designed to have the least resistance
strident: loud, harsh, and grating to the ears
sub- (prefix): underneath, below, less than
subcellular: of less than cellular scope or level of organisation
subfossil: the remains of an organism that are partially fossilised, meaning they are neither fully modern nor completely mineralised like a true fossil
submerge: completely cover or sink below the surface
submissive, submission: the act of accepting control by another
subordinate: a person or thing that has a lower rank or is less important than another
subpopulation: specific small section within a larger group
subspecies: additional species within a grouping
substrate: the material or substance an organism lives on or attaches to, or it's the molecule that an enzyme acts upon in a chemical reaction
subtropical: bordering on the tropics; nearly tropical. pertaining to or occurring in a region between tropical and temperate; semitropical
superficial (anatomy): on the surface
superior (anatomy): above, higher than
supra- (prefix): above, beyond, or earlier in time
supralittoral: area above shoreline, likely damp
survive, survival: stay alive, staying alive
susceptibility: easily affected, harmed, or influenced by something
swim bladder: special organ inside a fish that helps it float and control its depth in the water
sygnathids: a large and diverse group of fishes, including seahorses, pipefishes, seadragons and pipehorses
symbiotic, symbiosis: living together in a mutually beneficial way
synchronise/d: cause two or more things to happen, move, or operate at exactly the same time and speed

tactic: action or manoeuvre designed to achieve a particular goal
tail streamers: tail streamers on birds are the elongated narrow tips of the tail feathers; elongated, specialised tail feathers that are often used for display and communication, particularly during mating rituals. They can be quite long, sometimes exceeding the bird's body length, and are often vibrant in colour or pattern
talon: the sharp, curved claw on the foot of a bird of prey, used for hunting and gripping prey
tassel, tasselled: fringed lobes of skin similar to tassels, which are made of multiple cords, usually hanging down
tawny: dark or dull yellowish-brown
temperate: moderate in respect to temperature
tendril: thin organs of (usually of plants) that wrap around other objects for support
terrestrial: land based, growing or living on land
territory, territories, territorial: a tract of land, somewhere that the animal recognises as their territory which they will protect if they are territorial
testes: an organ which produces spermatozoa (male reproductive cells)
tetrodotoxin: an incredibly potent, deadly neurotoxin found naturally in certain marine and amphibian animals
thamnasteroid: refers to a specific skeletal growth pattern in colonial corals where individual polyp cups (corallites) lack defined walls and share their internal dividing walls (septa) equally with their neighbours
thermal:using, resulting from, or producing heat
thermal current: energy that is transferred from one body to another because of a difference in temperature creating a current
thicket: a dense growth of bushes or small trees that are packed closely together, often making it difficult to pass through
tholichthys: highly specialised, pelagic larval stage found only in certain fish families
thorax: the portion of the body between the head and the abdomen
threat: alarm, danger, hazard, menace, peril, risk, warning
threatened: in danger or likely to be harmed, or to become more significantly vulnerable or endangered
tidal: anything related to, caused by, or affected by **tides**
timid: shy, nervous, and lacking confidence, or it can mean overly cautious and hesitant due to fear
tissue: a group of similar cells and their intercellular material that work together to perform a function
tolerate: put up with
totem, totem animal: a natural object or animal that is believed by a particular society to have spiritual significance and that is adopted by it as an emblem
toxic/toxin/toxucity: poison, poisonous
trait: a specific characteristic or quality of an animal, or thing that makes it unique
translocation: the movement of something from one place to another
transmit, transmitted: to send or pass something from one place or person to another, like a message or a disease
transparent: see-through, light passes through completely like clear glass
trans-oceanic: crossing, spanning, or being situated on the other side of an ocean
transverse (anatomy): dividing body into upper and lower sections
trauma, traumatic: frightening, dangerous, or violent event that poses a threat to life or bodily integrity; wound or injury which can be both physical and/or psychological, or both
trenches (ocean zone): hadal zone

tropics, tropical: regions of Earth closest to the imaginary line called the equator, which runs around the middle of the planet. It's always warm in the tropics, with high humidity and plenty of rain

truncate: to cut off short; abbreviate, cut, the movement of something from one place to another, foreshorten, shorten

trypanosome: genus of **kinetoplastids**: a monophyletic group of unicellular parasitic flagellate protozoa

tubercles, tuberculate: a little bump or knob, kind of like a tiny wart, that can grow on plants or on the body, especially on bones

tube, tubular: long, hollow cylinder

tunicate/s: marine invertebrates (animals without a backbone) that belong to the subphylum *Tunicata* (sometimes called Urochordata). Often referred to as "sea s quirts", they are named for their unique, tough, and rubbery outer layer known as a "tunic," which acts as an exoskeleton

turbidity: cloudiness or haziness of a fluid caused by tiny suspended particles floating in it

turbulent/turbulence: characterised by conflict, disorder, or rapid, uneven changes. It can describe situations, periods of time, or even physical movements like air or water flows. Think of it as the opposite of smooth, stable, or calm.

tympanum: the eardrum or tympanic membrane

ultraviolet light: a type of electromagnetic radiation with a wavelength shorter than visible light but longer than X-rays, making it invisible to the human eye

undulate, undulating: to move or be shaped in a smooth, wave-like pattern

unicellular: consisting of a single cell

unique, uniqueness: one of a kind, unlike anything else

urban: relating to a particular place, town, village, in this instance development due to humans needing to have more homes

uterus: a hollow, muscular, pear-shaped organ in the female pelvis where a baby develops during pregnancy

vacuum: a space where most of the air or matter has been removed, creating a void

vagrant: wandering from place to place

vary, variable, variation: the quality or state of being different or diverse, not uniform

vegetarian: eating only plant matter

vegetation: plants

venom, venomous: poison, poisonous

venom glands: the specialised glands responsible for producing the poisonous secretions of an animal, often associated with spines, teeth, stings, or other piercing devices

vent (anatomy): the external opening of the **cloaca**, a single posterior orifice that serves as the only opening for the digestive, urinary, and reproductive tracts in many animals, including birds, reptiles, amphibians, and some fish and mammals

ventral (anatomy): towards the front

ventrolateral, ventrolaterally: the front and side of a body or body part

verify, verified, verification: prove the truth of, evidence

vertebra, vertebrae, vertebrate: spinal column or backbone

vertical: upright, top to bottom

vibrant: bright, dynamic, and pulsing with activity or strong, glowing colour.

vibrate/vibratory: to move back and forth or up and down rapidly and rhythmically, like shaking or quivering

vicinity: the area around something

victor: a conqueror or winner

vigorous/vigorously: showing a lot of energy, strength

virus: an infectious microbe consisting of a segment of nucleic acid (either DNA or RNA) surrounded by a protein coat. A virus cannot replicate alone; instead, it must infect cells and use components of the host cell to make copies of itself

vivid: strikingly bright or intense

vocalisation/s: the act of producing sounds with the voice

void: empty space

voracious: having a huge appetite; ravenous

vortex: a mass of whirling fluid or air that spins rapidly around a centre

vulnerable: weak, unsafe, defenceless, sensitive

wader/s: type of bird that typically lives near water and forages by walking in shallow areas

wary/wariness: cautious and alert to potential danger

water hole: a natural puddle or small lake in a dry area where animals go to drink

water-logged: saturate with water

water table: the boundary between saturated and unsaturated ground

weed: plant growing where it's not wanted

weir/s: bank, barrier, ditch, dike

wet sclerophyll forest: characterised by very tall eucalypt trees usually in moist cloudy uplands with up to 2500mm of rainfall each year

wetlands: an area of land that's covered by water or has waterlogged soil, meaning the ground is full of water, like a sponge

whorl/s: something that swirls or winds around like a spiral

wingspan: the maximum extent across the wings of a bird or other flying animal, measured from tip to tip

xanthophore/s: a **chromatophore** containing a yellow pigment that is typically a **carotinoid** and occurring especially in fishes and crustaceans

zooplankton: tiny, animal-like organisms that drift or float in water

CORAL POLYP STRUCTURE (SCLERACTINIANS)

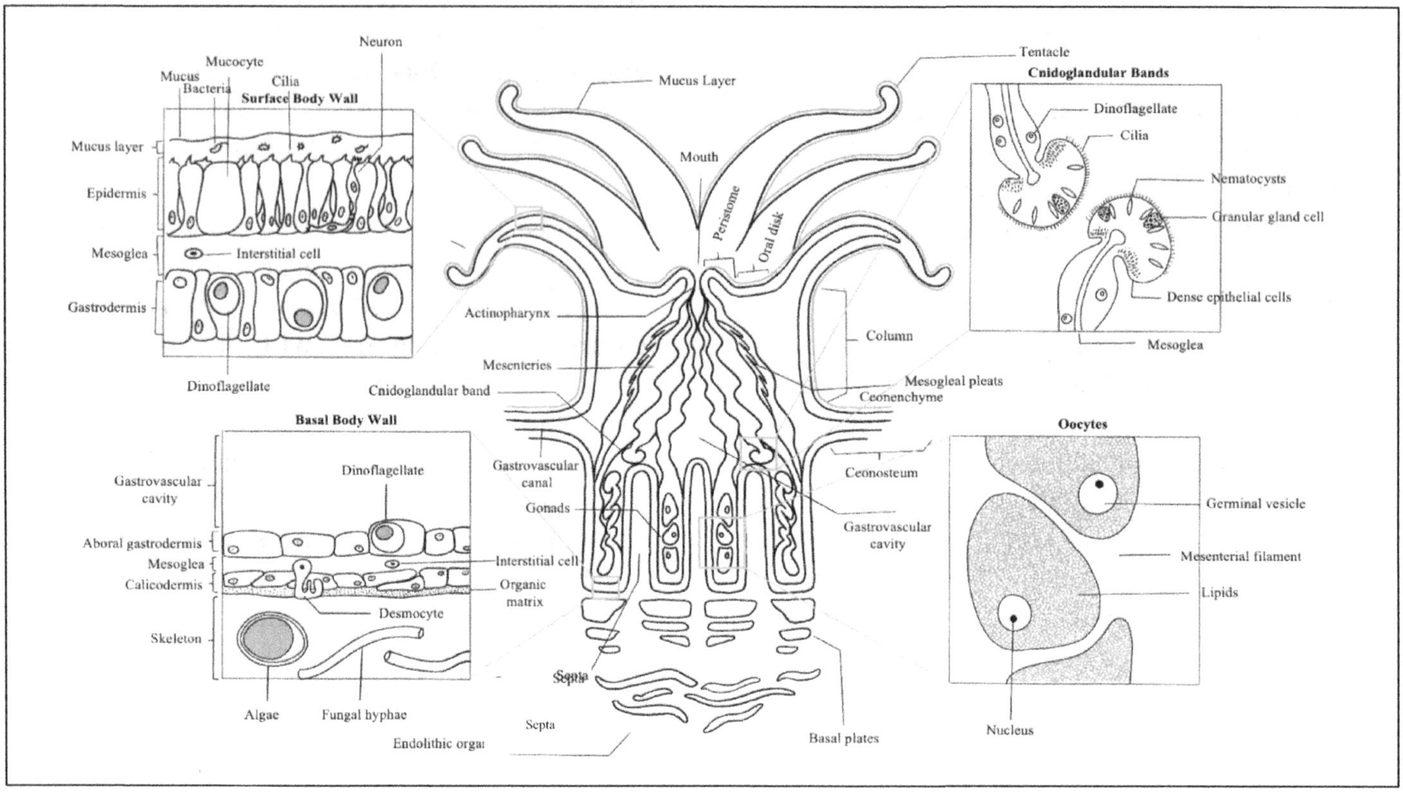

Image credit: Murphy McDonald, NOAA

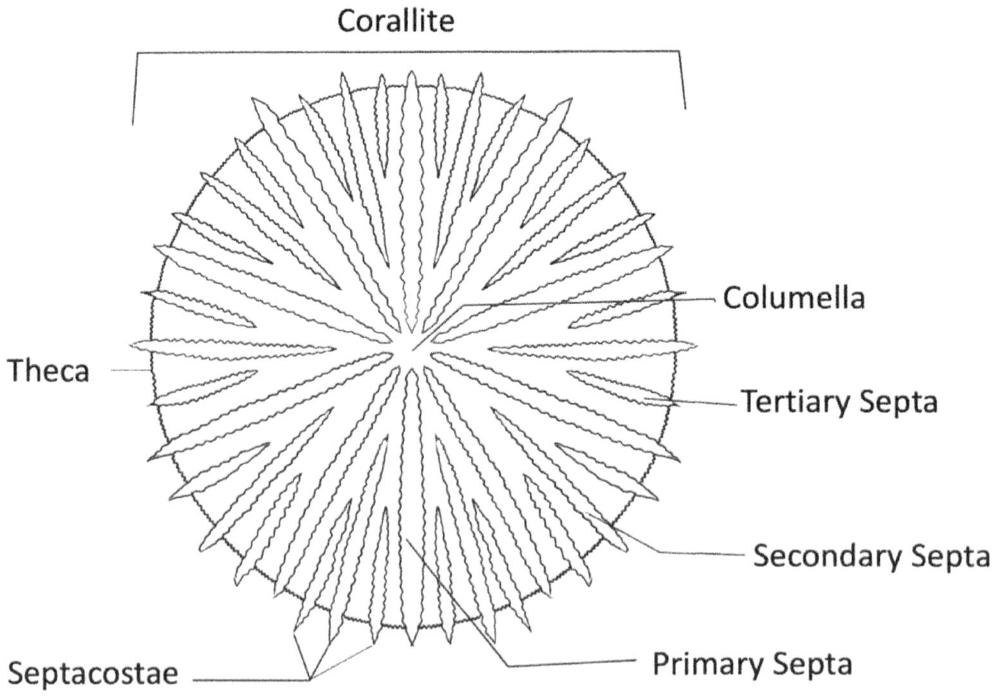

Image credit: Murphy McDonald, NOAA

To claim your Conservation Club reward badge for THE GREAT BARRIER REEF MARINE PARK environment:

- **Answer the following 20 questions.**
- **Send either a photograph or a scan of one colouring page** which you have coloured in reasonably correctly as far as the colours of your chosen species are concerned!

*** Postage will apply both in Australia and overseas. This will be estimated and sent to you for approval via email prior to shipping.**

Let's see what you have learned, whether you are 8, 80, or somewhere in between!

The answers can be found throughout the book.

1. What are the babies, born line, of hammerhead sharks called?
2. One name for a group of rays is a school – what is the other?
3. What species are the largest shorebirds in Australia?
4. Which species is known as the only non-human species to create musical instruments?
5. What type of fish is "Dory" from the movie 'Finding Nemo'?
6. Which bird has a specialised velvet black plumage that makes them look like a black hole in the rainforest?
7. Which fish is the largest of the wrasse species and is listed as Endangered on the IUCN Red List?
8. Which marine creature has the largest brain of any fish species?
9. Which tiny fish is named after the Norse god of mischief?
10. Which species of fish can swim backwards?
11. Which fish can tie its own tail into a knot and slide the knot down its body to enable it to tear its prey apart?
12. What is a group of bar-tailed godwits called?
13. Which mammal is known for its complex, haunting songs?
14. Which bird is famous for its mating dance, and also has hybridised with the Sarus crane?
15. Which fish's teeth turn a vivid, electric blue as the fish matures?
16. Which fish has one of the fastest strikes in the animal kingdom, as fast as 0.006 seconds?
17. Which shorebird is one of the heaviest shorebird species in the world weighing around 1kg?
18. Which fish was named after a Sesame Street character and only officially described as a new species in 2026?
19. Which fish can quickly change colour from a uniform bright yellow to mottled grey, dark green or brown?
20. Which mammal is known for its acrobatic displays being able to spin lengthwise up to seven times in a single jump?

ABOUT THE AUTHOR

Meet Elysia Rainwood, an artist and writer with a passion for conservation and environmental protection, particularly the preservation of the diverse animals inhabiting our wonderful Australian continent.

Elysia developed a passion for the unique creatures and the diversity of plant life found in Australia early in life. Elysia always had an eye for detail, being very aware of such things as seasonal changes in the various environments observed. From the depths of the lush, green rainforest to the sun-kissed coral cays and lagoons of the Great Barrier Reef, her adventures fueled a desire to share the stories of the vulnerable and endangered.

As an adult, Elysia worked part-time as a "reef interpreter" in the Capricorn Bunker Group of islands, primarily Northwest Island and occasionally Tryon Island; this involved taking day visitors on reef walks, island walks, glass-bottom boat rides and guided snorkelling experiences, and talking about the Great Barrier Reef and some of the creatures to be found there. She also enjoyed regular family camping trips to Northwest Island and Lady Musgrave Island with her young children and, in fact, clearly recalls donning a child-carrier backpack with her son for reef walks when he was still a toddler. Therefore, this book is particularly dear to her heart!

Elysia writes non-fiction books and stories that resonate with the spirit of the tiniest of creatures and soar with the majesty of the albatross. She likes to focus on lesser-known species that may be in your backyard, ensuring they and their habitats are not overlooked.

Elysia Rainwood's mission is clear:
To enchant young minds with the wonders of Australia's wildlife, fostering a generation of conservationists. Through interesting facts and engaging, reasonably accurate illustrations, she hopes to ignite a spark in the hearts of 8- to 14-year-old readers, inspiring them to protect the precious creatures that share our planet. Of course, many older readers are also enthralled by the quality of information and illustrations, so Elysia's audience is significantly wider than was originally intended!

Along the way, Elysia hopes to also engage with parents, caregivers, educators who can pass on this information, as well as those who just enjoy the relaxation that colouring careful illustrations can bring.

So, whether you are a youngster or someone who shares this passion for conservation, join Elysia on this informative journey into the magical realms of nature, where words become wings for the endangered and vulnerable to soar.

INDEX

Made in the USA
Monee, IL
08 November 2026

67968595R00090